Table of Contents

This table of contents includes material contained in this manual as well as software and documents on the accompanying CD.

Preface v

Acknowledgments viii

Installation and Operation ix

Disclaimer x

1. One-Dimensional, Steady-State Conduction

    1.1 Project: One-Dimensional, Steady-State Conduction in Composite Systems 1

    1.2 Extended Surface Heat Transfer 9
        a. HttExtnd.exe (CD)
        b. Appendix - Modeling Details (pdf, CD)
        c. Appendix - Tridiagonal Systems of Equations (pdf, CD)
        d. SGTSV - (CD)

2. Two-Dimensional, Steady-State Conduction

    2.1 Two-Dimensional, Steady-State Conduction 14
        a. Htt_2dss.exe (CD)
        b. Twoddss.xls (CD)

3. One-Dimensional, Transient Conduction

    3.1 One-Dimensional, Transient Conduction 22
        a. HttOnedt.exe (CD)
        b. Transient Conduction in a Semi-Infinite Slab (semicalc3.xls, CD)

    3.2 Project: Monticello Problem 31

    3.3 Project: Sandwich Wall Construction 37
        a. Sandwich Wall Slideshow (pdf, CD)

    3.4 Project: Transient Conduction at the Interface between Two Materials 42

4. Forced Convection - External Flows

    4.1 Forced Convection on a Flat Plate 46
        a. HttFlatp.exe (CD)
        b. Appendix - Laminar, Forced Convection on a Flat Plate Algorithm (pdf, CD)
        c. Appendix - Self-Similar, Boundary Layer Flows (pdf, CD)
        d. Rungek.for (CD)
        e. Rungek90.for (CD)

4.2 Project: Convective Heat and Mass Transfer from a Runner 56
      a. Water and Air Properties (airwaterprops.xls, CD)

5. Forced Convection - Internal Flows

   5.1 Forced Convection - Internal Flows 63
      a. Htt_Pipe.exe (CD)
      b. Appendix - Laminar, Thermal Entry Length Heat Transfer (pdf, CD)
      c. Appendix - Turbulent Pipe Flow (pdf, CD)

   5.2 Internal Flow Correlations 72
      a. Corrinternal.xls (CD)

6. Natural Convection

   6.1 Natural Convection in a Saturated Porous Layer 75
      a. HttPorus.exe (CD)
      b. Appendix - Implementation Using Vorticity-Streamfunction (pdf, CD)
      c. Appendix - Implementation Using Primitive Variables (pdf, CD)

   6.2 Benard Convection Using Primitive Variables (pdf, CD)

7. Heat Exchangers

   7.1 One-Dimensional Heat Exchangers..81
      a. Htt_HX1D.exe (CD)
      b. Appendix - HX1D Algorithm (CD)

   7.2 Two-Dimensional Heat Exchangers..86
      a. Htt_HX2D.exe (CD)
      b. Appendix - HX2D Algorithm (CD)

8. Radiative Heat Transfer

   8.1 Project: Transmissivity of Glass 90
      a. Planck's Law and Transmissivity Data (plnkslaw.xls, CD)

   8.2 Calculation of Radiation View Factors by Nusselt Unit Sphere Method 94
      b. Htt_View.exe (CD)

   8.3 View Factor Spreadsheets and Charts 99
      a. Viewfactors2000.xls (CD)
      b. Viewfactor Charts (pdf, CD)

9. Appendix - An Excel/Visual Basic for Applications (VBA) Programming Primer (pdf, CD)

   9.1 Demosub.xls (CD)

Preface

The seeds of this project were planted a number of years ago while I was working as a research engineer in the nuclear industry. Back then we used punched cards, and during my innumerable hikes hauling them back and forth to the computer center, I had plenty of time to ponder the state of computer use in that industry. Frankly, it made me uncomfortable. I had had the benefit of a good computational fluid dynamics and heat transfer course myself in graduate school and had done mostly computational modeling in my dissertation research. Even with that as a background, I felt overwhelmed by the volume and range of the modeling efforts going on around me. That was two decades ago. Since then, modeling, simulation and visualization have moved from an activity practiced mostly by Ph.D.s in certain "high-tech" pockets of industry to an activity involving just about any engineer involved in research, design and manufacturing. Just look at the dozens of advertisements for various modeling and analysis packages advertised in the professional magazines!

As a result of that early experience, I have tried to integrate as much modeling and simulation as I can into all the courses I teach. In addition to the implementations of the algorithms themselves, the visualization, verification and interpretation of computed results have always been major components of any project I assign. Indeed for many of the projects contained on the CD we implemented the graphical display before the program itself was running in order to aid in development and debugging.

The projects described in this manual and software contained on the accompanying CD-ROM have been developed over nearly 15 years and incorporate what I have learned myself and tried to teach in a half dozen different courses, both graduate and undergraduate. A few were begun almost immediately when we got our first PCs in the mid 1980s. Some were taking much of their current form as early as 1987 when they were first used in a televised, graduate course in Computational Fluid Dynamics (CFD). In a "methods" course like CFD, a simple, graphically-rich demonstration can often make a point much more vividly than the corresponding and often obtuse differential equation or a page full of numbers. Originally True Basic and later Watfor-77, both of which included graphics primitives and were easy to program and debug, were used. In those earlier versions of this software, the calculations were slow and the graphics quite crude. "Multitasking" was often necessary - that is, you would start the solution of a parabolic equation, and it might have crunched on to the downstream end by the time you got back from lunch!

By 1990 we built our first local classroom with a projection system, and with that the pace of development quickened. By now it was possible, for instance, to actually demonstrate in our on-campus classes the effect of parameters like the Reynolds and Prandtl numbers on the growth of boundary layers and not just ask students to take it on faith! And when something seemed amiss or if it appeared that many students didn't understand the graphical presentation, we fixed it immediately. The project got a big boost during the 1995-96 academic year when the present Visual Basic interfaces were developed. At that point the programs could be used not just by the instructor for in-class demonstrations as before, but also by the students themselves. As a result we switched our undergraduate heat transfer survey course to a two-lecture-a-week format supplemented by a two-hour-long "studio session." These hands-on sessions are held in a room equipped with a computer for each pair of students and are run very much as one would a real physical lab. In the six years of the studio sessions, we have continued to enhance and improve this software.

While we cannot guarantee the accuracy of the entire collection contained on the CD, there is nothing like a team of 60 or so students simultaneously using a piece of software to find its inadequacies. Thus, since the initial roll-out, much effort has gone into correcting and enhancing these modules. This work included not only correcting actual errors, but changing the display of the results where the graphical presentation was not as illustrative as it could be. In one case (the view factor module), the whole algorithm was scrapped and replaced with another when students found erroneous results for certain ranges of parameters. For almost every module of the nine contained herein, there are cases for which a benchmark solution is available (often an analytical solution right out of any standard text), and we make it a point to always include such a comparison exercise with the use of each module. I hope that any new user will do likewise.

Since most commercial CFD packages can be used on many of the problems these modules were developed to handle, one might ask why we did not just use one. Most educators who have tried that option report having had to divert too much class time to "learning the software." That drawback was exactly the reason we developed a custom interface for each module. Learning this software is synonymous with learning the heat transfer! That is, there are no extraneous inputs and no extraneous outputs. If the user doesn't know the significance of a Fourier number in transient conduction or can't do a simple steady-state energy balance on a control volume and isn't willing to learn, then this software will not help him or her much!

The write-ups contained in this manual for the nine modules include at least some physical background on the computations taking place behind the scenes, and all write-ups include operating instructions for the software. Each module also includes a "Help" file containing similar operating instructions. Several are very complete. In many cases a separate write-up discussing the algorithm used is available as a "pdf" file on the CD-ROM. These write-ups, which were prepared as the algorithms were being implemented as class assignments and projects in a number of different courses, are meant to be detailed enough that the interested user could replicate the major features of the software. These descriptions are not necessarily original; indeed several are my attempt to fill in the missing steps in the description of an algorithm given elsewhere so that my students would be able to implement it in a reasonable time. The implementation of several of these algorithms makes an excellent week-long or several week-long project in a graduate heat transfer or CFD course.

In most cases the actual problems and exercises to be done using the module have not been included here, but are announced on a class web page. In the past we have used many problems directly out of standard textbooks. Creating new exercises, especially design projects, which more thoroughly exploit the speed and visualization capabilities of this software is an ongoing project, and we do welcome any feedback from users. Indeed, once one sees how much several of these modules, e.g., the two for heat exchangers, eliminate the tedium of conventional problem solving, one soon recognizes that much more of the available class time can and should be devoted to engineering design. Similarly, the physical experiments we have developed at my university to accompany several of these modules are only mentioned in passing in this manual.

In addition to the nine "canned" modules, a few other projects that are likely to be implemented on a spreadsheet have been included in the manual. These projects were initially developed because we didn't have enough of the VB/Fortran modules to fill all 14 studio sessions of the semester. Several have proven as educationally useful as the canned modules. Several other spreadsheets, e.g., implementing an analytical solution, are also included in the package and at least described in this document. A number of the latter were developed expressly to aid in verification of the numerical models emphasized here.

Acknowledgments

I would like to acknowledge a number of people who and programs that have contributed directly or indirectly to this project.

Many of the predecessors of these modules were developed beginning as early as 1987 to be used as demonstrations and assignments in several courses taught by way of TV through Virginia's Commonwealth Graduate Engineering Program (CGEP). Through this innovative, multi-university consortium, working professionals throughout the state and beyond may earn a Master of Engineering degree through distance education. The expectation of these students that an engineering course of the late 20th century would use modern pedagogical tools – as they use modern engineering tools in their jobs – was powerful incentive for developing and continually improving these programs. The support of CGEP and the encouragement by my off-campus students are gratefully acknowledged.

Mr. Jerry O'Leary, who was supported at first by the University of Virginia Teaching + Technology Initiative (T+TI) and then continued working on his own time, designed and implemented eight Visual Basic interfaces. This includes not just the graphical user interfaces (GUI's) visible on the screen, but also the interaction between the VB and the underlying Fortran "engines." He continues to provide advice freely as these modules have been improved and updated over the past several years.

Dr. Polley McClure, until recently Vice President for Information Technology and Communication and Chief Information Officer of the University of Virginia and now holding a similar position at Cornell University, provided moral and financial support through the T+TI.

Mrs. Harriet C. Weed funded the Lucien Carr III Chair in Engineering Education at the University of Virginia, which I held from 1992 to 1995. A number of the Fortran engines behind these modules were developed with that support.

Dr. Glen Bull of the Curry School of Education at UVa allowed me to audit two of his graduate courses in instructional technology. What I learned from Glen – particularly that you shouldn't spend the time trying to make a computer do what has already been done well in print – is heavily reflected in this work.

Graduate students in MAE 611 (Heat Transfer), MAE 712 (Convection Heat Transfer), MAE 672 (Computational Fluid Dynamics and Heat Transfer) and undergraduates in ME 320 (Computer Graphics) have been helping to develop the algorithms used in these modules beginning in the early 1980's and continuing to the present. Members of the UVa Mechanical Engineering Classes of 1997 – 2002 have helped immeasurably in debugging both the software and this manual.

Finally, I would like to thank Jonathan Plant, senior editor at McGraw-Hill, whose encouragement over the last few years has resulted in this work finally getting to print!

Installation and Operation

This software collection comes with an automatic "install" program. However, if you would like to install it manually, the details are given in this section.

SYSTEM REQUIREMENTS

The programs in the Heat Transfer Tools collection are personal computer applications designed to run under Microsoft Windows. They have been successfully tested under Windows 3.1, Windows 3.11, Windows 95, Windows 98 and Windows NT. They will run best on a PC 486 or better with at least 8 MB of random access memory. The display requires a screen resolution of at least 800x600 (with small fonts) in order to view the entire graphical user interface. A mouse is required to perform most operational functions. While the Visual Basic programs included in this collection will operate only under Microsoft Windows, the Excel workbooks may also work on Macintosh computers.

INSTALLATION INSTRUCTIONS - Visual Basic 3 Programs

Eight of these modules consist of an executable (.exe) developed using Microsoft Visual Basic 3.0 plus a collection of subroutines written in Fortran and compiled as a single Windows dynamic link library (.dll). Windows Help files (.hlp) are included for each of these eight. Sample data files (.ss2) are also included for one module (Htt_2DSS). As an example, the files making up the one-dimensional transient conduction module are as follows:

   HttONEDT.EXE - the main executable
   HttONEDZ.DLL - the dynamic link library of Fortran subroutines
   HttONEDT.HLP - the help file written specifically for this program

Certain Microsoft system files are needed in order to run the VB-3 programs. These files are contained in the root directory of the CD, and if you are running the programs from the CD, no further action may be required. If you intend to install the software on a hard drive and run them from there, the following three files are to be installed in your WINDOWS or WINDOWS\SYSTEM directory:

   VBRUN300.DLL
   CMDIALOG.VBX
   THREED.VBX

Whether you run from the CD or from a hard drive, the Microsoft True Type "Fences" fonts must be installed in your Windows system in order to read and print certain equations given in the Help topics of the HttONEDT and Htt_2dss programs. The FONTS portion of the Windows Control Panel can be used to "install" these fonts. The FENCES.TTF file is included in the root directory on the CD. Unless these fonts are installed properly, certain bizarre symbols will appear instead of the intended parentheses, brackets and braces.

To install the collection of modules on your harddrive:

1. If Microsoft Visual Basic 3.0 is not already installed on your PC, copy the three Visual Basic libraries to your \WINDOWS or \WINDOWS\SYSTEM directory.

2. Install fonts

3. Create a directory on your hard drive: e.g., \HTT

4. Copy the application-specific files to that directory. The programs run slightly faster from a hard drive.

5. Create icons for each module. Make sure the specified Working Directory is the same as the directory containing the executable and its associated Fortran DLL.

INSTALLATION INSTRUCTIONS - Visual Basic 6 Program

The most recently developed program, HttPorus, is written completely in Visual Basic Version 6, which unlike VB-3 is a compiled language and thus fast enough for fairly intensive calculations. That being the case, only a single file is associated with this particular module. If the appropriate systems files happened to have been installed on your platform by some other application, then the HttPorus program can be launched by simply double-clicking on the file name. If you get an error message citing missing files, then it will be necessary to install those system files on your hard drive. These are all contained in the HttPorus.CAB file which may be unpacked by running the accompanying setup.exe file.

INSTRUCTIONS - Excel Workbooks

Six Excel workbooks (files ending in ".xls") are included on the CD. Several of these were developed to help in the debugging of the VB-3 modules; others are intended to support one or more student projects. To use them you must have a recent version of MS Excel. All six contain macros written in Visual Basic for Applications. Upon opening each you may receive a warning about the possibility of those macros containing viruses. Be aware that these workbooks could themselves become infected from your previously infected installation.

INSTRUCTIONS - "pdf" Files

Some fifteen files, many of them detailed descriptions of the algorithms embedded in these programs, are provided on the CD in "pdf" form. Providing your computer already has the Adobe Acrobat reader installed, these files may be viewed and printed if desired by double clicking on the file name. The Adobe Acrobat reader may be downloaded free at: http://www.adobe.com/products/acrobat/readstep.html.

DISCLAIMER

THESE PROGRAMS WERE CREATED FOR INSTRUCTIONAL PURPOSES ONLY. THEY ARE PROVIDED AS-IS WITHOUT WARRANTY OF ANY KIND, AND THE MCGRAW-HILL COMPANIES, INC., AND THE AUTHOR EXPRESSLY DISCLAIM ALL IMPLIED AND EXPRESS WARRANTIES, INCLUDING, BUT NOT LIMITED TO, THE IMPLIED WARRANTIES OF MERCHANTABILITY AND FITNESS FOR A PARTICULAR PURPOSE. THE MCGRAW-HILL COMPANIES, INC., AND THE AUTHOR DO NOT WARRANT, GUARANTEE OR MAKE ANY REPRESENTATIONS REGARDING THE USE OR THE RESULTS OF THE USE OF THESE PROGRAMS.

THE MCGRAW-HILL COMPANIES, INC., AND THE AUTHOR SHALL NOT BE LIABLE TO ANYONE FOR ANY INACCURACY, ERROR OR OMISSION IN THESE PROGRAMS, REGARDLESS OF CAUSE, OR FOR ANY DAMAGES RESULTING THEREFROM.

IN NO EVENT WILL EITHER THE MCGRAW-HILL COMPANIES, INC., OR THE AUTHOR BE LIABLE TO ANY THIRD PARTY FOR ANY INCIDENTAL OR CONSEQUENTIAL DAMAGES (INCLUDING, WITHOUT LIMITATION, INDIRECT, SPECIAL, PUNITIVE OR EXEMPLARY DAMAGES) ARISING OUT OF THE USE OF OR INABILITY TO USE THESE PROGRAMS OR FOR ANY CLAIM BY ANY OTHER PARTY, EVEN IF SUCH PARTY HAS BEEN ADVISED OF THE POSSIBILITY OF SUCH DAMAGES.

1.1 One-Dimensional, Steady-State Conduction in Composite Systems

Introduction

In this project we will study a system consisting of multiple thermal resistance elements in a general series/parallel arrangement. A configuration consisting of just four elements is seen in Figure 1 below, where the predominant heat flow is in the horizontal direction. An everyday example is that of a typical residential stud wall construction in which wooden studs every 16" or 24" alternate with an insulating material such as fiberglass. These materials are generally sandwiched between sheathing, drywall, siding, etc. to form a series-parallel arrangement (which is somewhat more complicated than that depicted in Figure 1). Assuming that the materials designated as "2" and "3" in Figure 1 below have different values of thermal conductivity, the problem is at the very least two-dimensional. The two methods discussed here involve analyzing this geometry as a 1-D problem, but between the two establish upper and lower bounds for an estimate of the overall thermal resistance (or conductance). If the computed bounds are tight enough to satisfy our purposes, we can avoid the more laborious (and maybe nearly impossible) task of a complete 2-D or 3-D analysis.

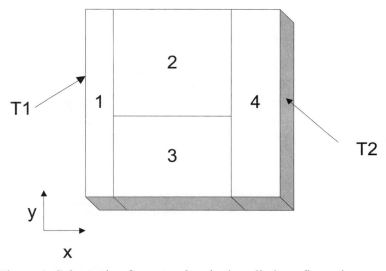

Figure 1. Schematic of a general series/paralled configuration

Analysis

Let us look at the geometry shown schematically in Figure 1 and consider two approaches to modeling it. (Bear in mind that a stud wall construction will have several more layers in series on one or both sides.) We will assume that each of the four elements shown can be approximated as a single resistor. Of course we are assuming steady-state conditions, no internal heat generation, constant thermal properties within each element, etc., so that a network of electrical resistors is a valid analog. There is assumed to be no contact resistance between the elements.

In estimating the effective thermal resistance of this four-element network, we might take either of two approaches. The first method assumes that isotherms are planes perpendicular to the x axis (which runs straight through our "sandwich"). In this case our equivalent circuit is as shown in Figure 2.

1.1 One-Dimensional, Steady-State Conduction in Composite Systems

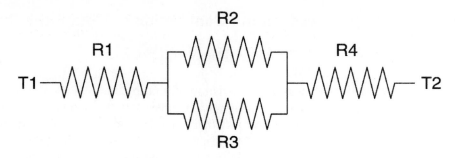

Figure 2. Planar isotherm model for configuration shown in Figure. 1

This "planar isotherm" approach will give us an upper bound on the conductance of this composite system (or alternatively a lower bound on the resistance). The reason is that with this model any heat passing through Region 1 can take whichever path it finds easier to get through Regions 2 and 3. So for instance, if the thermal conductivity of Region 3 greatly exceeds that of Region 2, then some of the heat passing through the upper part of Region 1 will be diverted down to take the easier path through Region 3. This phenomenon is sometimes referred to as "thermal bridging." With the planar isotherm model, we are assuming that the transverse conductance is infinite (corresponding to zero transverse resistance).

The other way we can model this system is to assume that heat flux lines are straight in the x direction. (See Fig. 3.) This "straight line heat flux" model gives a lower bound on conductance (upper on resistance) because in this model any heat going through Region 1a (the upper part of the first region), for instance, is forced to continue straight through Region 2 (and Region 4a), even though Region 3 may offer less resistance to heat transfer.

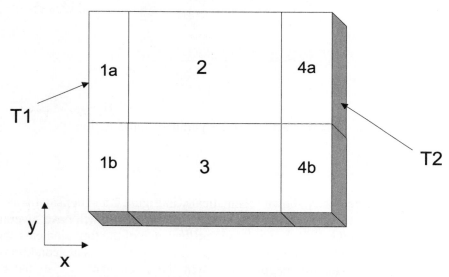

Figure 3. Schematic for the straight line heat flux model

1.1 One-Dimensional, Steady-State Conduction in Composite Systems

The equivalent circuit for this model is shown below. Here we assume the transverse conductance is zero (infinite resistance) so that heat is forced to follow a path straight through, even though it may have to pass through regions of higher resistance.

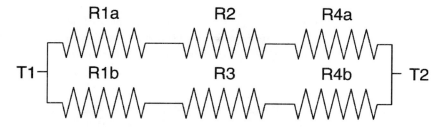

Figure 4. Equivalent circuit for straight line heat flux model of Figure 3

As long as the frontal area (A), thickness (L) and thermal conductivity (k) of each of the elements above are available, it is easy to figure the effective thermal conductivity of each circuit element (from $R_{cond} = L / Ak$ for a slab) and, using them, the effective resistance (or conductance) for the multi-resistor circuit (Figs. 2 and 4). With the terminal temperatures T_1 and T_2 known, one can in turn find the overall heat transfer $q = \dfrac{T_1 - T_2}{R_{total}}$ and any intermediate temperatures.

A More Difficult Calculation

Let us make this problem somewhat more difficult by specifying, as seen in Figure 5, convective and radiative boundary conditions at one side rather than a fixed temperature.

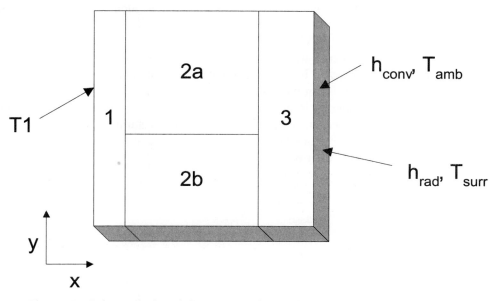

Figure 5. Schematic involving convective and radiative boundary condition in parallel on one side, fixed temperature at other end

1.1 One-Dimensional, Steady-State Conduction in Composite Systems

Now the heat flux out the right side is given by:

$$q = q_{conv} + q_{rad} = h_{conv}\left(T_{surf} - T_{amb}\right) + h_{rad}\left(T_{surf} - T_{surr}\right).$$

Note that T_{surr} is a temperature representing that of the solid surfaces with which the surface exchanges radiant energy, while T_{amb} is representative of the surrounding air and may well be different. Using the planar isotherm model, an equivalent circuit for this configuration is shown in Figure 6.

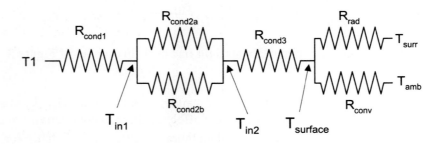

Figure 6. Equivalent circuit corresponding to Figure 5

Here $R_{rad} = \dfrac{1}{h_{rad} A}$ and, even with a number of simplifying assumptions, is a function of the surface emissivity (ε), $T_{surface}$ and a temperature T_{surr} which is representative of the surrounding surfaces with which it exchanges radiant energy. The derivation of the radiation heat transfer coefficient is given in the appendix. In a similar fashion $R_{conv} = \dfrac{1}{h_{conv} A}$ where h_{conv} is a function of $T_{surface}$, $T_{ambient}$, geometric parameters and properties of the fluid in which this system is assumed immersed. We assume a state of natural convection; that calculation is also discussed in the appendix.

If T_{surr} and T_{amb} were the same and, moreover, the radiative and convective heat transfer coefficients were known, then it would be straightforward to calculate an equivalent resistance for the circuit in Figure 6. Since that is not the case, a reasonable approach is to implement an iterative solution, which will gradually improve the values of the two surface resistances while finding the unknown interior temperatures T_{in1}, T_{in2} and $T_{surface}$. Assuming that the two surface resistances are updated continuously, then heat balances applicable at the three nodes where the temperatures are unknown provide the necessary equations:

$$\frac{T_1 - T_{in1}}{R_{cond1}} + \frac{T_{in2} - T_{in1}}{R_{cond2}} = 0$$

$$\frac{T_{in1} - T_{in2}}{R_{cond2}} + \frac{T_{surface} - T_{in2}}{R_{cond3}} = 0$$

$$\frac{T_{in2} - T_{surface}}{R_{cond3}} + \frac{\left(T_{ambient} - T_{surface}\right)}{R_{conv}} + \frac{\left(T_{surr} - T_{surface}\right)}{R_{rad}} = 0$$

1.1 One-Dimensional, Steady-State Conduction in Composite Systems

These three heat balance equations may be solved for T_{in1}, T_{in2} and $T_{surface}$, respectively and used as predictive equations for those same quantities. The iterative sequence consists of updating the two surface resistances using the current value of $T_{surface}$ and then updating each of the three unknown temperatures. This sequence is repeated until the change in all quantities is negligible from iteration to iteration. In the implementation represented by the spreadsheet on the next page, the iteration is handled in a custom Visual Basic for Applications (VBA) subroutine invoked whenever the "Runit" button is pressed. This iteration could also be accomplished using the built-in "Solver" routine in Excel^{tm}. Columns have been set up for the temperatures themselves as well as others for inputting and computing resistances between the interfaces. Cells having a white background are intended for user input. The solid layers were deliberately specified to be very thin so that conductive resistances were in turn small compared to the surface convective and radiative resistances. That disparity in the relative size of the resistances is clearly evident in the plot of the resulting temperature distribution through the layer.

References

1. Çengel, Yunus A., *Heat Transfer, A Practical Approach*, WCB McGraw-Hill, NY, 1998.

2. Incropera, F.P. and DeWitt, D.P., *Fundamentals of Heat and Mass Transfer*, 4th Ed., Wiley, NY, 1996.

3. Gottfried, B.S., *Spreadsheet Tools for Engineers Excel 97 Edition*, WCB McGraw-Hill, NY, 1998.

1.1 One-Dimensional, Steady-State Conduction in Composite Systems

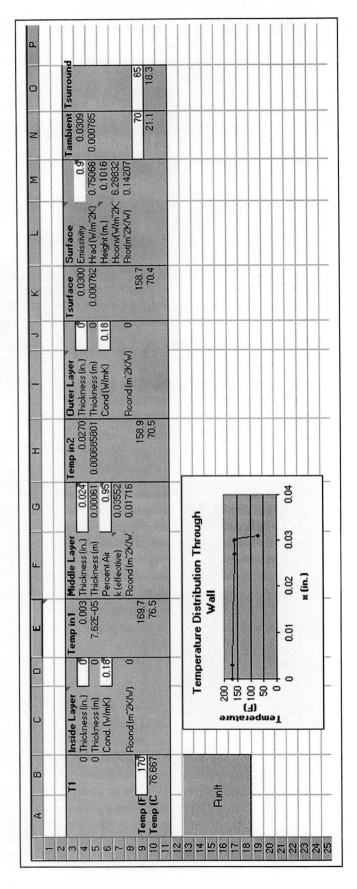

Figure 7. Spreadsheet for calculation discussed in last section. A VBA subroutine implementing an iterative procedure to find the three unknown internal temperatures and the surface convective and radiative heat transfer coefficients is invoked when the button is pressed.

1.1 One-Dimensional, Steady-State Conduction in Composite Systems

Appendix: Calculation of the Radiative and Convective Heat Transfer Coefficients

We consider here only the simple case of a small, convex object having emissivity ε at a temperature of $T_{surface}$ exchanging heat with much larger surroundings all at the same temperature T_{surr}. The radiative heat exchange is then given by:

$$q''_{rad} = \varepsilon \sigma \left(T^4_{surface} - T^4_{surr} \right)$$

The temperature term may be factored to yield:

$$q''_{rad} = \varepsilon \sigma \left(T^2_{surface} + T^2_{surr} \right)\left(T_{surface} + T_{surr} \right)\left(T_{surface} - T_{surr} \right)$$

The first four factors on the right hand side then comprise the radiative heat transfer coefficient, h_{rad}. Note that unlike convection heat transfer coefficients, which are not usually strong functions of temperature, the radiative heat transfer coefficient definitely is. With the iterative solution proposed in this project, it may be kept current easily as the surface temperature is improved. A simple Visual Basic for Applications (VBA) function that computes the radiative heat transfer coefficient is shown below:

```
Function Hrad(T1 As Single, T2 As Single, Epsilon As Single) As Single
'Computes radiation heat transfer coefficient.
'Temperatures T1 and T2 are assumed to be passed in degrees Celsius
'and must be converted to degrees Kelvin
'Epsilon is the emissivity of the surface.
'This is Eqn. 3-11 from Cengel's Heat Transfer - A Practical Approach
'WCB McGraw-Hill, 1998

Dim T1K As Single, T2K As Single, Sigma As Single

Sigma = 0.0000000567
T1K = T1 + 273
T2K = T2 + 273
Hrad = Epsilon * Sigma * (T1K ^ 2 + T2K ^ 2) * (T1K + T2K)

End Function
```

For purposes of this calculation, we assume that the convection on the outside surface is buoyancy-induced and given by Churchill and Chu's correlation for laminar natural convection on a vertical surface:

$$\overline{Nu}_L = 0.68 + \frac{0.670 \, Ra_L^{1/4}}{\left[1 + (0.492/Pr)^{9/16} \right]^{4/9}}.$$

Here the Rayleigh number is given by: $Ra_L = \dfrac{g\beta(T_s - T_\infty)L^3}{\nu \alpha}$, where L is the characteristic length scale (height). A VBA function implementing this correlation is on the next page.

1.1 One-Dimensional, Steady-State Conduction in Composite Systems

Within that function other VBA functions are invoked for the necessary properties of air. This property library is discussed in the forced convection from a runner problem.

```
Function HNatConv(Tsurf As Single, Tambient As Single, Height As Single) As Single

'Computes convective heat transfer coefficient for laminar, natural
'convection on a vertical, flat plate using Churchill and Chu's correlation
'(Eqn. 9.27 in Incropera and DeWitt's, Fundamentals of Heat and Mass
'Transfer, 4th Ed., Wiley).  It is only good to a Rayleigh number of 10^9,
'but that should be sufficient here.

Dim Tfilm As Single, Gravity As Single, Rayleigh As Single
Dim Prandtl As Single, Bottom As Single, NuBar As Single

Tfilm = 273 + 0.5 * (Tsurf + Tambient)
Gravity = 9.8

Bottom = (Viscosity_Air(Tfilm) * Conductivity_Air(Tfilm)) _
    / (Density_Air(Tfilm) ^ 2 * Cp_Air(Tfilm))

Prandtl = Prandtl_Air(Tfilm)
Rayleigh = Gravity * (1# / Tfilm) * (Tsurf - Tambient) * Height ^ 3 / Bottom
If Rayleigh > 10 ^ 9 Then
        MsgBox "Rayleigh number out of range!"
        Exit Function
End If
NuBar = 0.68 + (0.67 * Rayleigh ^ 0.25) / (1 + (0.492 / Prandtl) ^ (9 / 16)) ^ (4 / 9)
HNatConv = Conductivity_Air(Tfilm) * NuBar / Height

End Function
```

Compos.doc 7/7/00

1.2 Extended Surface Heat Transfer

Introduction

This module deals with heat transfer from extended surfaces – the thin protuberances often referred to as "cooling fins." Specially designed arrays of extended surfaces as are commonly used for cooling electronic equipment are known as "heat sinks." The rationale for using extended surfaces may be seen by looking at the *convective* resistance, as used in the DC electrical analogy to conductive heat transfer:

$$R_{conv} = 1/hA \quad \left[= \frac{\Delta T}{Q} \right]$$

Fins are used typically when the convective heat transfer coefficient (h) is relatively small and cannot easily be enhanced. In order to compensate, the designer increases the area (A) exposed to the ambient fluid. At the same time they must be careful not to increase the conductive resistance in series between the heat source (e.g., the cylinder wall in an air-cooled aircraft piston engine or a power transistor in an electronic device) and this surface convective resistance. Thus we generally find that fins are thin, are made of a highly conductive material like copper or aluminum and are used when the ambient medium is a gas, most commonly air. By working with this module, you will be able to test and see readily the effects of the various design parameters, including geometry, thermal conductivity and convective heat transfer coefficient.

A very common example of use of extended surfaces is the inexpensive baseboard convector shown in Figure 1. (Devices like these are often loosely referred to as "radiators.") The equivalent circuit for this configuration <u>without the fins</u> is shown in Figure 2 (with any radiative loss to ambient ignored). The three resistances shown correspond to: (1) the convective resistance inside the pipe: $R_{conv-in} = \dfrac{1}{h_i 2\pi r_{inside} L}$, (2) the convective resistance at the outside surface: $R_{conv-out} = \dfrac{1}{h_o 2\pi r_{outside} L}$ and (3), the conductive resistance of the pipe itself:

$$R_{cond} = \frac{\ln \dfrac{r_{outside}}{r_{inside}}}{2\pi L k}.$$

Figure 1. Baseboard convector

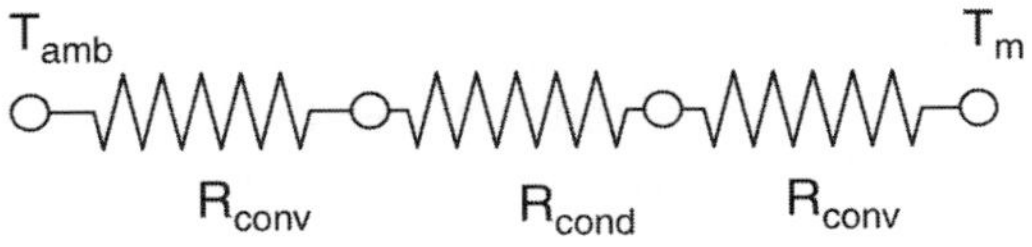

Figure 2. Equivalent circuit for Figure 1 configuration

1.2 Extended Surface Heat Transfer

If we use typical values of $h_i = 1000$ W/m^{2}K, which is characteristic of forced convection of water, and $h_o = 10$ W/m^{2}K, which is characteristic of natural convection in air with dimensions corresponding to the unfinned copper pipe in Figure 1 (outside radius = .011024 m, inside radius = .010516 m), we find that the convective resistance on the outside of the pipe is about 95 times the convective resistance inside (the reason it is not 100% corresponding to the 100 fold difference in h is the slightly larger area) while that resistance is in turn over 800 times greater than the conductive resistance of the thin, highly conductive copper (k = 401 W/mK) wall. (Actual computed values are $R_{Conv-inside}$ = .0151 K/W, R_{cond} = 1.87x10^{-5} K/W and $R_{Conv-outside}$ = 1.4427 K/W.) Here where our design calls for transferring as much heat as possible to the room, the high thermal resistance corresponding to the low natural convective heat transfer coefficient in air is clearly the bottleneck. Note that we have ignored any thermal resistance associated with less than perfect contact between the fin base and the copper pipe. This application will be revisited later when we apply results from this module.

Analytical solutions of the governing energy balance equation are available for many fin geometries, but here we use a *numerical* procedure based on the *finite-volume* method. Our numerical simulation accomplishes the same thing as the analytical techniques covered in most texts, but does it at a more fundamental level. Succinctly put, we assume, as is customary in extended surface analysis, that a one-dimensional conduction model is applicable. Here one-dimensional means that conduction is *primarily* along the fin axis direction. Clearly there must also be conduction in the lateral directions to get the heat out to the surface where it can be transferred to the fluid.

To obtain a numerical solution, we "chop" (discretize) the length of the fin into a reasonable number of segments or volumes and write a statement of conservation of energy for each. This conservation statement includes conduction into the small volume from its higher temperature next-door neighbor, convection from its exposed surface to the adjacent fluid and conduction out to its other, lower temperature neighbor. There is no storage of energy within the volume and no volumetric generation. But instead of shrinking the representative control volumes to infinitesimal length (as we would do in deriving the governing ordinary differential equation), we allow the control volumes to remain of finite size. Then we approximate the derivatives arising from the application of Fourier's law in the two conduction terms by changes in temperature over small distances. Similarly we apply Newton's Law of Cooling to model convection losses from the surface. From the representative energy conservation statement, we discover that the temperature of each of these numerous cells depends only on itself and its two closest neighbors, and thus a tridiagonal system of algebraic equations results. Complete details of the algorithm are given in the appendix.

This *numerical* solution is more powerful in many respects than an analytical solution. As long as we can express the appropriate cross-sectional areas for conduction (at the interfaces between our small volumes) and corresponding exposed surface areas for convection, such an algorithm can handle any 1-D geometry readily. Moreover such an algorithm could handle non-uniform thermal conductivity or position- or temperature-dependent convective heat transfer coefficient (although neither capability is currently included in this module). A numerical solution such as this does not reveal the functional form of the resulting solution, e.g., that the temperature distribution may be expressed in terms of a combination of hyperbolic sines and cosines or Bessel functions. But it will give the numerical values of the temperature for whatever parameters have been specified and will allow the user to "see" the entire temperature distribution. Except in the case of straight fins of uniform cross-section where the functional form of the analytical solution is fairly simple, the temperature distribution is almost never computed. Evaluating the analytical solution for, say, an annular fin is no picnic since it involves

1.2 Extended Surface Heat Transfer

Bessel functions - that is why you will never see it plotted; all you will find is a bottom-line design parameter, the fin efficiency plotted as a function of a non-dimensional fin parameter. But by seeing the full temperature distribution, as you will here, you will see immediately why a particular design has high or low fin efficiency.

For this module the numerical work has already been done and is provided in the form of a Visual Basic/Fortran application program. This computational model should give the same results similar to those one gets from using the equations and graphs, and one should solve a few problems both ways. **All** programs should be checked as well as possible; in this case the standard analytical solutions provided in any heat transfer text provide convenient benchmarks. Once you are familiar with the operation of the module and are sure you can rely on the results, you can run many cases easily in order to familiarize yourself with the physical processes involved.

Program Operation

To begin the user must select the type of fin to be analyzed. When activated the **Select Fin** button opens another form that allows the user to select one of four common geometries: straight rectangular, cylindrical pin, annular and triangular fins. A schematic of that particular geometry is displayed (See Figure 3 below) whenever one of these configurations is selected. The triangular fin is assumed infinite in extent in the third direction, while the straight rectangular fin is of finite width. (You can handle a 2-D (infinite) rectangular fin by making the third dimension (W) very large relative to the others.) The necessary dimensions for all four geometries are clearly indicated on each diagram.

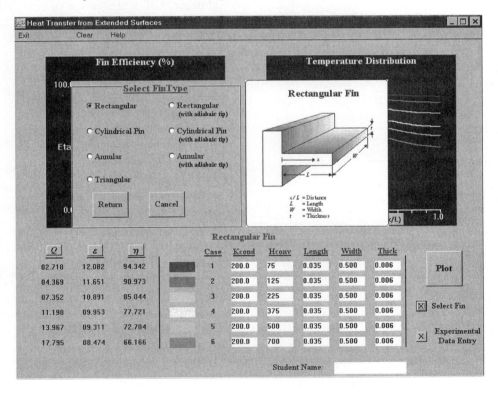

Figure 3. Interface with "Select Fin" form displayed

1.2 Extended Surface Heat Transfer

In addition to the four common geometries, the first three, but with an adiabatic tip, may also be selected. This capability may be useful for certain problems, including extended surfaces involving symmetry. Once the fin type has been selected the user may enter input data (in the white squares at the bottom right) for as many as six fins of that single type. Required data includes the dimensions appropriate to the particular geometry selected, as well as the thermal conductivity of the material and the surface convective heat transfer coefficient.

A typical output screen is seen in Figure 4. The temperature distribution along the fin is shown in the upper right window for as many as six fins. Note that the scaling length for each fin is its own length. Thus if you are testing a family of fins of different physical length, the temperature distribution plots will all still run from $x = 0.0$ to $x = 1.0$. Also with the temperature scaling used, the base (root) temperature is always 1.0, while for a very long fin the tip temperature will approach 0.0. The fin efficiency is plotted as a function of the fin parameter in the upper left window. The latter may be checked readily against the usual textbook charts, where the same parameter is used as the abscissa. A fin with little temperature drop along its length will show a high efficiency, and similarly a fin whose temperature drops nearly to that of the fluid in just a fraction of its length will register a low efficiency. Both plots are color keyed to the six input cases. In addition the total heat transferred, fin efficiency and fin effectiveness (which is less often used in design than the fin efficiency) values are returned in tabular form in

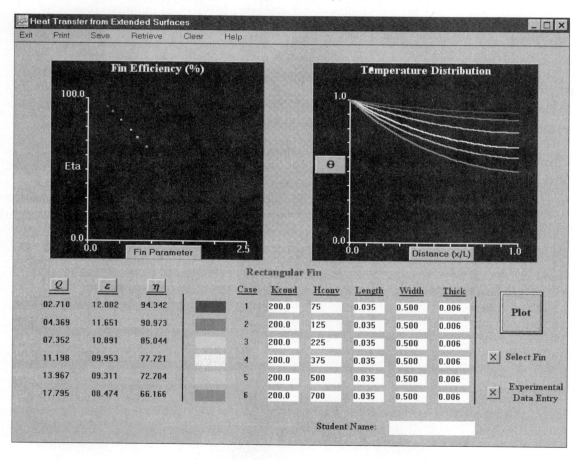

Figure 4. Output screen

1.2 Extended Surface Heat Transfer

the lower left. Cases resulting in low effectiveness values (which indicate that the 1-D conduction model used here is not appropriate) are flagged. "Hot buttons" at the top of each column provide concise definitions of these outputs.

The symbols seen in the plot on the right hand panel are experimental values of temperature which may be entered into the **Experimental Data Entry** subform in normalized form for comparison with numerical predictions. Again the position is normalized by the actual fin length, while temperatures are normalized by the difference between the base and fluid temperatures.

An Application: The Baseboard Convector

The module does not provide for geometry identical to that seen in Figure 1, so we will approximate the square fins as annular. With an inner and outer radii .011024 and .032243 m. respectively, and a thickness of .000203 m, the module returns values for a single aluminum (k = 177 W/mK) annular fin of Q = .051 W/K, effectiveness (ε) = 361.7 and an efficiency = 87.5%. Clearly the total heat transfer per fin (Q) can be found from the efficiency ($Q = h\, A_{fin}\, \eta$) as well as from the effectiveness ($Q = h\, A_{base}\, \varepsilon$). This Q (W/K), however computed, must be multiplied by the actual temperature difference between the base temperature and the ambient fluid temperature to find the actual heat transfer (W).

In the introduction we computed the resistances corresponding to a 1.0 m length of the copper pipe seen in Figure 1. Thus for a valid comparison, we must consider a configuration consisting of certain percentage of that length being unfinned and the rest finned. With 177 fins per meter and with each fin .000203 m thick, we find that 3.6% of the length is finned, while 96.4% is unfinned. Assuming, as is usually done, that the same value of convective heat transfer coefficient applies both to the finned and to the unfinned area, then:

$$Q_{total} = h * (.964 * 2\pi R_{pipe} + 177 * A_{fin}\eta)$$

Now we find Q_{total} = 9.67 W/K or R = 1/9.67 = .10343 K/W, compared to the thermal resistance of 1.4427 K/W found for the bare pipe. This low-tech installation of fins has cut the worst of the three resistances in the baseboard convector by a factor of 14 and the total resistance by a factor of 12!

Program Verification

Efficiency values for the four basic fin geometries may be easily compared with those given in the conventional charts of all textbooks and should match well.

Reference

Çengel, Y.A., *Heat Transfer - A Practical Approach,* McGraw-Hill, NY (1998), pp 287 – 290.

Extended.doc 6/12/00

2.1 Two-Dimensional, Steady-State Conduction

Introduction

In this module we study two-dimensional, steady-state conduction using the *finite-difference/finite-volume* method, a numerical technique that involves the discretization (chopping up) of the solution domain into a finite number of cells. Unlike analytical solutions which, once determined, can be evaluated at any point in the domain, we will be satisfied here with an approximation of the temperature at a relatively small number, i.e., dozens, hundreds, or maybe several thousands, of points. The temperature computed at each of these points is taken to be representative of the small volume surrounding it. Two sample problems, both of which have analytical solutions involving infinite series, have been provided so that the user can study and understand the input scheme used for the numerical solution. The program will also evaluate the analytical solution for those same two cases and allow the user to compare it with the numerical solution and to investigate the effect of truncating the series after a particular number of terms.

The user is expected to derive the heat balance equations for any particular nodal configuration (internal cells, edges, corners, etc. with a fixed temperature, specified heat flux or convection situation specified) and to input that information to the program. The user is in effect providing the coefficients for a large, sparse system of linear equations (one equation for each of the cells into which the domain has been divided). The module performs an iterative solution of the resulting system of equations and then depicts the computed solution in the form of a color contour plot of isotherms.

It is the responsibility of the user to interpret and verify the validity of that contour plot in the context of the problem being solved. In other words, the user, not the program, does the heat transfer! All that the software does is solve the equations quickly and then display the computed results. It is the user's job to see to it that all the inputs are correct and that what is eventually plotted on the screen is, in fact, a valid solution to the problem at hand. Many modern commercial heat transfer software packages have the various possible thermal boundary conditions available for the user to select with the click of a "button" or through an input data file. Even though a setup like that would have been easier for our programmer than what is implemented here, we have avoided going to that extreme of "user-friendliness." The practicing engineer in industry is assumed to already know what the user of this module is only now learning and to be able to use software intelligently! That is why we have required the user of this module to derive the heat balances. By the time you are done with this exercise, you should certainly know how to perform a heat balance on any type of cell.

It is well to keep in mind that what you are setting up mathematically on the computer is exactly analogous to a DC electrical circuit analog. Such analogs were used frequently before the advent of the digital computer, especially when an analytical solution was not possible. In both cases we model a physical system that is *continuous* by a *lumped* system. The only real differences are that your computer model is a lot easier to modify than the electrical mockup (no soldering required!) and the numerical results are slightly easier to measure and display than the experimental results. On the other hand, with the electrical mockup you actually "see" the configuration you are working on right in front of you. The corner of one such analog built to model a two-dimensional cooling fin is shown in Figure 1. Each of the wood screws sticking further out represents a finite-difference cell. Each is connected to its two, three or four neighbors; this represents the conduction path through the metal of the fin itself. The thick copper wires running along the plywood base correspond to the ambient air to which this cooling

2.1 Two-Dimensional, Steady-State Conduction

fin is exposed. The other resistors running seemingly randomly between the taller woodscrews and the heavy copper leads running along the base represent the convective resistance of each cell. As is the case with the computer model, half cells are used along edges and quarter cells at corners. Noting that the conductive resistance for a slab is expressed as R= L/Ak, then with half the conduction area, as is the case along edges, the resistance is doubled. This explains the multiple resistors seen in these areas.

Figure 1. Detail of electrical analog of two-dimensional cooling fin. Each of the screws sticking further out corresponds to a finite-difference (volume) cell; the heavy lines (ground) correspond to the ambient air temperature

Background

Very extensive on-line help files accompany this module, so only a cursory survey is given in this section. The two-dimensional steady-state program includes an input screen where information is entered for the particular problem at hand (see Fig. 2). Setting up that input is how you will spend much of your time. When a partial differential equation is solved numerically, the solution domain can be divided into a number of zones where a particular "prescription" applies. The information you enter on the input screen for each zone tells the program the range (in both directions) of that particular zone and what prescription to apply to each cell in that zone. For instance, both sample cases consist mostly of generic internal points, plus edges and corners where other predictive equations are used. In general these special equations come from the application of boundary conditions and depending on which type of boundary condition is applied (fixed temperature, specified heat flux, convection) may be easy or more difficult to derive.

The derivation of the predictive equations for internal and special boundary nodes is dealt with at length in the on-line help files; so here we will look at a generic internal node only briefly. Consider first a typical internal point with no volumetric heating, no convection normal to the cell, constant thermal conductivity and grid spacing.

2.1 Two-Dimensional, Steady-State Conduction

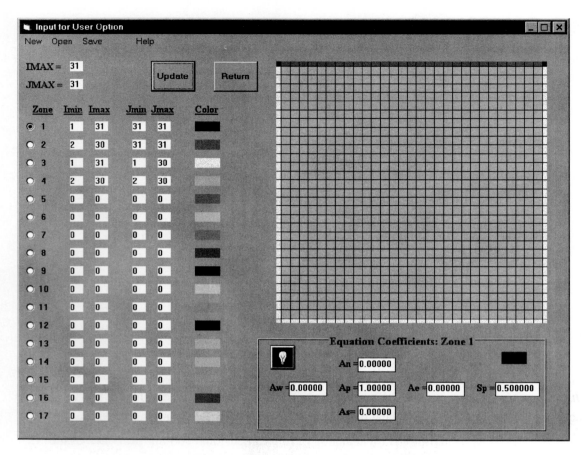

Figure 2. Setup form corresponding to solution shown in Figure 5 below

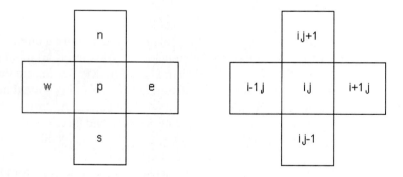

Figure 3. Five-point finite-difference "stencil." Compass notation (left) and i,j notation (right)

In steady-state we can write the heat balance equation in terms of the fluxes into and out of the four faces of the center control volume (See Fig. 3):

$$-k\Delta y \frac{T_p - T_w}{\Delta x} - (-)k\Delta y \frac{T_e - T_p}{\Delta x} - k\Delta x \frac{T_p - T_s}{\Delta y} - (-)k\Delta x \frac{T_n - T_p}{\Delta y} = 0.0 \quad (1)$$

or equivalently,

2.1 Two-Dimensional, Steady-State Conduction

$$k\Delta y \frac{T_w - T_p}{\Delta x} + k\Delta y \frac{T_e - T_p}{\Delta x} + k\Delta x \frac{T_s - T_p}{\Delta y} + k\Delta x \frac{T_n - T_p}{\Delta y} = 0.0 \qquad (2)$$

(Unit depth in the third direction (into the paper) is assumed.) In Equations 1 and 2 we have used the more illustrative "compass" notation to indicate the point (p) and its four neighbors (n, s, e, and w); the actual code works in the more conventional (and vital for the computer) "i,j" subscript notation. (One of the minus (-) signs in each of the four terms in Equation 1 comes from Fourier's Law: $q = -kA\frac{dT}{dn}$. The convention that a positive heat flux through the bottom or left side of a cell is a net gain to that cell; while a positive heat flux through the top or right hand side is a loss explains the other signs.) For $\Delta x = \Delta y$ and uniform thermal conductivity (k), Equation (2) can be simplified to give an equation of the form:

$$a_p T_p + a_w T_w + a_e T_e + a_s T_s + a_n T_n = S_p \qquad (3)$$

As long as we restrict ourselves to the five-point "stencil" of Figure 3, then it turns out that the governing equation for <u>any</u> point can be put into the form of Eqn. 3. Thus we have used that equation as the entire basis of our program.

For the case resulting in Eqn. 2 above, $S_p = 0.0$; when there is volumetric generation as in the second sample, S_p will have a non-zero value. A non-zero S_p could also come from the hAT_∞ part of a surface convection term or from a prescribed surface heat flux. Observe that this equation comes directly from a heat balance at node p and if solved for T_p, gives us a "prescription" for that value in terms of its four neighbors (which, of course, are usually unknowns as well) and possibly a "source" term. The coefficient values we would load into the program for all internal points are then $a_p = -4.0$, $a_w = a_e = a_s = a_n = 1.0$ and we would tell the program to apply this particular prescription at all points other than those along the four boundaries.

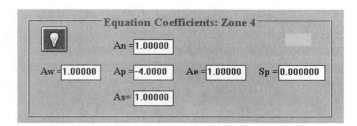

Figure 4. Coefficients and right hand side of Eqn. 3 for internal points in first sample

Handling fixed temperature boundary conditions is very easy. For the first sample, where the temperatures on all four sides are prescribed, all the neighboring coefficients (a_w, a_e, etc.) for boundary points are identically 0.0, the coefficient for the points themselves (a_p) is 1.0 and the right hand side (S_p) is whatever the prescribed value is (here 0.0 on the sides and bottom, 1.0 along the top). At the upper corners, where the temperature is discontinuous, we might apply the mean of the side and top values, namely 0.50. This value really does not enter into the calculation of unknown values, but its effect can be seen in the contour plot locally.

2.1 Two-Dimensional, Steady-State Conduction

The fact that we derived Equations 1 and 2 directly from an energy balance on a small, but finite-sized cell makes what we are implementing here technically the *finite-volume* method. The older *finite-difference* method implies that we start from the governing partial differential equation (PDE) and represent (in the case of the steady-state conduction equation) the second derivatives by differences in temperature over small increments. For the simple case just derived, the end result is the same. For more complicated geometries such as cylindrical coordinates and for cases involving variable properties, etc., the finite-volume method is preferred.

If you are starting a problem from scratch or studying one of the samples, it is highly advisable that you draw the region to scale on gridded paper, number nodes as necessary, and label each of the zones for which a different prescription applies. (You might want to think in terms of working these problems on an electronic spreadsheet in which case you would enter a cell formula once and then, because you certainly don't want to fill in each cell formula separately, copy it to the whole range for which it applies.) Often all the internal cells make up a single zone. Usually each of the boundaries and sometimes corners must be treated separately as a zone requiring its own set of coefficients. It is best to start with a clean worksheet (Figure 2) by clicking on **New,** then filling in the appropriate ranges for each of your designated zones in succession. Note that zone specifications further down the list override those closer to the top. In other words you could have Zone 1 encompass the whole computing region, and then cut out other regions as needed further down.

Determining the coefficients a_p, a_e, etc., and the right hand side S_p may be trivial as in the case of a fixed temperature (Dirichlet) condition or may require you doing a detailed energy balance on a representative cell (as will be done later in the discussion of one of the example problems). The program is currently set up for a maximum of seventeen zones (only four are used in the sample shown in Figure 2). Since running a program on successively finer meshes is an important part of verifying a solution, you might want to make your diagram as general as possible. For instance, you might label the largest value of the horizontal index as I = Imax on your diagram, rather than say, I = 20.

Once you tell the program to run, it will set up a system of Fortran Do Loops and run until it has converged to a solution. An outer loop controls the overall calculation, stopping when the changes from sweep to sweep at all points have been reduced to within specified (and very small) limits. This may take anywhere from maybe 50 to several hundred sweeps, but in any case won't take long. The inner Do Loops range over the various regions. One set of doubly-nested loops might cycle over all general interior points; another single loop might cycle just along a horizontal boundary where a particular prescription is applied. This prescription might be trivial, as in the case of a fixed walled temperature, or it might be more complicated as would arise when a convection condition is specified at a boundary. The iterative technique used in this program is called the Modified Strongly Implicit (MSI) procedure [1]. Like the Gauss-Seidel iterative method usually discussed in texts [2], MSI involves multiple sweeps over all the equations, gradually improving the solution with each sweep. Although adequate for most textbook problems, Gauss-Seidel tends to be very slow for large sets of equations. MSI, being about 30 years newer and significantly faster, was chosen from perhaps a dozen possible algorithms for implementation in our program. With this fast and efficient solver and use of dynamic array allocation, the size of your mesh is limited only by the available memory of your computer.

While deriving and inputting the limits of zones and the numerical prescriptions are the responsibility of the user, certain safeguards have been built into the pre-processing section of the program. The "Scarborough" criterion [3] says that for an iterative solution of the sort used here to converge, the absolute value of the coefficient of the visited point (a_p) must be greater than or equal to the sum of the absolute values of the coefficients of the neighbors (a_e, a_w, a_n, and a_s) at all points, and, furthermore, must exceed that sum at least at one point. The former is automatically satisfied if you do the energy balances correctly, while the latter ensures appropriate boundary conditions are applied somewhere. If this built-in safeguard is violated, you will get the warning message: "This problem will not converge! Check Diagonal Dominance." At this point the calculation ceases awaiting your changes. But there is no guarantee that a converged solution is the correct answer to your own problem. Only your diligent effort at deriving and inputting the coefficients in the predictive equations and your active verification of the results can ensure an accurate solution!

The potential user probably wonders why this module might be preferable to any of the commonly available electronic spreadsheets as a vehicle for solving Laplace and Poisson equations (or any elliptic equation) numerically. There are at least four good reasons. For a start, the color coding on the input sheet makes the various zones readily apparent (although one could certainly change the text or background color on a spreadsheet to indicate the ranges of the different cell formulae being used). The several checks of input data built into the setup routine are not there in a spreadsheet. (Try applying the adiabatic condition on all faces with uniform volumetric generation on a spreadsheet and see what happens.) Also the iterative scheme incorporated into most spreadsheets is rudimentary and on large ("non-toy") problems would run much slower than the MSI algorithm used here. Finally, this module provides the two analytical solutions (next section) already embedded within it for study by the user.

Sample Problems

Two sample problems, both having analytical solutions, have been provided with this module, and you are encouraged to study both their setup sheets and the solutions. The first of these involves a square region subject to one fixed temperature on three boundaries and a third different, fixed temperature on the fourth. The input sheet for this problem is depicted in Figure 2. (To get this sheet yourself, select **User-Defined Problem** and **Return,** if necessary. Then on the input sheet select **Open** from the menu and choose the file **sample1.ss2.**) Zone 1 has been selected in Figure 2 as the active zone so that in the bottom right one sees the coefficients and right hand side corresponding to the fixed boundary temperature (0.5) along the entire upper boundary. Zone 2 then resets all these points other than the two corners to the actual boundary temperature of 1.0. (Note that subsequently defined zones override those closer to the top.) When Zone 4, which corresponds to the 29x29 internal points, is selected, one sees the numbers seen in Figure 4 as derived above.

The analytical solution for this problem is discussed in great detail in Section 4.2 of Incropera and DeWitt [2] and in many other sources. In Figure 5, isotherms based on the numerical solution are shown as colored bands, while white lines superimposed on the color plot show isotherms computed on the same grid, but based on the analytical solution. In the solution shown in Figure 5 five terms have been included in the summation; distinct discrepancies are apparent near the top of the plot. The user is encouraged to experiment with the number of terms included. This same solution implemented in Excel and VBA, its macro language, and displayed more dramatically as raised, color contour plots is available on the CD.

2.1 Two-Dimensional, Steady-State Conduction

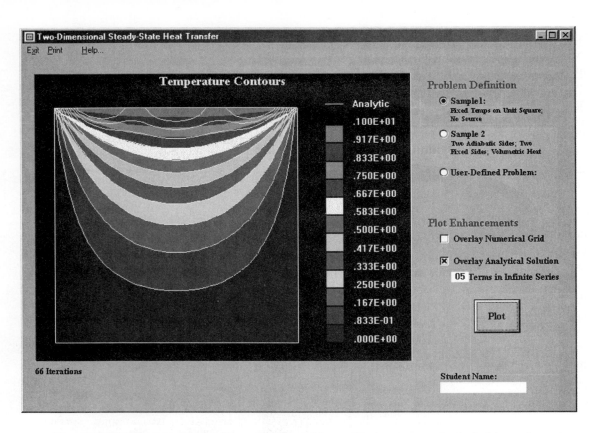

Figure 5. Output screen for first sample problem. Five terms are included in series.

The second sample (sample2.ss2) includes uniform volumetric heat generation in a unit square. Because of symmetry we need only compute the solution in the first quadrant. That scenario gives two fixed-temperature and two adiabatic boundaries. You ought to be able to derive any of the cell heat balances, including those for both internal and edge cells. Note that, as in the first sample, half cells are used along edges and quarter cells at corners.

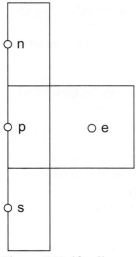

Figure 6 Half cell along left boundary

The energy balance for a typical half cell along the left side of this quadrant would look like this:

$$h \Delta y (T_\infty - T_p) - k \frac{\Delta x}{2} \frac{T_p - T_s}{\Delta y} - (-) k \Delta y \frac{T_e - T_p}{\Delta x}$$

$$-(-) k \frac{\Delta x}{2} \frac{T_n - T_p}{\Delta y} + \dot{q}''' \frac{\Delta x}{2} \Delta y = 0.0$$

If we now take h = 0.0, then the left side will be adiabatic. If in addition we take $\frac{\dot{q}'''}{k} = 1.0$ and $\Delta x = \Delta y$ and then multiply through by 2.0 we get: $-4T_p + T_s + 2T_e + T_n = -\Delta x \Delta y$. With the grid spacing used ($\Delta x = \Delta y = 1/40$), we then find S_p = -.000625.

2.1 Two-Dimensional, Steady-State Conduction

The resulting coefficients and right hand side for the adiabatic boundary along the left side are shown in Figure 7. Note that heat balance for these cells does not involve a neighbor to the west (there is none) and that the non-zero S_p is because of the volumetric generation. If instead this same left side were subjected to a convective boundary condition, the known fluid temperature would wind up on the right hand side (in Sp) so that the Aw coefficient would still be 0.0

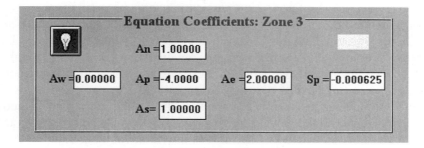

Figure 7. Coefficients and right hand side of Eqn. 3 for nodes along the left (adiabatic) side in second sample problem

This problem also has an analytical solution that is described in the help files [4], and the user can experiment with how many terms to include. In this second sample, where there are no discontinuities, only a few terms are needed to yield good agreement between numerical and analytical solutions. Observe carefully the behavior of isotherms at the two adiabatic surfaces. You should be able to explain why the isotherms are packed closer together near the two isothermal exterior surfaces. You can probably do a back-of-the-envelope calculation to verify that the maximum temperature computed is reasonable.

References

1. Schneider, G.E. and Zedan, M., "A Modified Strongly Implicit Procedure for the Numerical Solution of Field Problems," *Numerical Heat Transfer*, 4, pp. 1-19, (1981).

2. Incropera, F.P. and DeWitt, D.P., *Fundamentals of Heat and Mass Transfer*, 4th ed., Wiley (1996).

3. Patankar, S.V., *Numerical Heat Transfer and Fluid Flow*, Hemisphere, New York (1980).

4. Carslaw, H.S., and Jaeger, J.C., *Conduction of Heat in Solids*, 2nd ed., Clarendon Press, Oxford (1959).

Twodss.doc 7/10/00

3.1 One-Dimensional, Transient Conduction

Introduction

This module calculates and displays the transient temperature distribution for three geometries: the infinite slab, the infinite cylinder and the sphere, each exposed to a sudden change in surface convective conditions. Initial and boundary conditions as well as material properties are assumed to be such that a one-dimensional analysis is appropriate. The calculated results are commonly presented in heat transfer textbooks in the form of "Heisler charts," which are graphical representations of analytical solutions obtained by separation of variables in the 1940's. In lieu of a computerized evaluation of these analytical solutions, this module instead performs a numerical solution of the governing conservation equations and presents the resulting temperature profile at the conclusion of each time step. In this approach we "discretize" the computational domain (in both time and space) and use a finite-volume scheme to arrive at a system of algebraic equations, which are then solved for the nodal temperatures. A graphical display of the temperature profiles is presented dynamically as the computations proceed toward a user-selected stopping criterion. While requiring only a few more inputs than the classical (and static) Heisler charts, this module allows users to witness the whole transient evolution in time and space.

Very standard numerical techniques have been employed; these are described thoroughly both in the on-line documentation and in Reference 1. Thus only a brief overview is given here. The conservation of energy requirement applied to the transient conditions of this module yields a single equation applicable to infinite slabs, infinite cylinders and spheres alike:

$$\rho c_p \frac{\partial T}{\partial t} = \frac{k}{r^p} \frac{\partial}{\partial r}\left(r^p \frac{\partial T}{\partial r}\right) \tag{1}$$

Here the exponent p takes the values 0, 1 and 2, respectively and all material properties including density (ρ), specific heat (c_p) and thermal conductivity (k) are assumed to be uniform and constant.

Non-dimensionalization of the above equation and of the boundary conditions yields two dimensionless parameters, including a non-dimensional time, the Fourier number:

$$Fo = \frac{\alpha t}{L_c^2}, \tag{2}$$

and the Biot number:

$$Bi \equiv \frac{hL_c}{k} \quad \text{or} \quad \frac{hr_o}{k}, \tag{3}$$

The latter is a measure of the ratio of internal conductive to external convective resistance of the body. For the infinite slab L_c is the half width; for the cylinder and sphere r_o is the radius. The thermal diffusivity $\alpha = \dfrac{k}{\rho c_p}$.

The solution and evaluation of this parabolic, initial value problem by separation of variables is non-trivial in all three cases [2, 3]. For this reason most textbooks include graphs of the solution as a function of Fourier and Biot numbers for uniform initial temperature with one

3.1 One-Dimensional, Transient Conduction

side adiabatic and the other exposed to convection, based on a one-term approximation to the infinite series [4].

To solve the governing equations numerically, we must first "discretize" the 1-D spatial domain into small finite increments. A sample discretization is shown in the Figure 1:

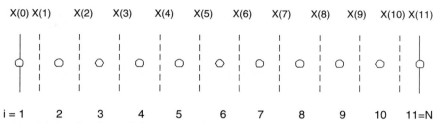

Figure 1. Discretization of the computing region (shown with 10 grid spacings only)

Here, for illustration purposes, the domain has been divided into nine equal increments, plus half-increments at the centerline ($i = 1$) and at the exposed surface ($i = N$). Temperatures are to be computed at the nodal locations indicated by the small circles in the figure. The vertical dashed lines mark the boundaries between computational cells. The X-locations indicate the distance measured from the line of symmetry for the infinite slab or from the center of the cylinder or the sphere. (In the actual module, we have selected $N = 21$, giving 19 equal increments plus 2 half-increments, or $\Delta x = 1/20$.)

The more traditional method of discretizing the governing equations is to replace the partial derivatives in Equation 1 with finite-difference approximations. We chose instead a more physically intuitive method based directly on conservation principles. If one derives the energy balance for a representative finite volume (a thin slab for the plane wall problem, a thin cylindrical shell and thin spherical shell for the infinite cylinder and the sphere, respectively), one recognizes that (after dividing through by the volume term) a single form is valid for all three geometries:

$$\frac{\theta_i' - \theta_i}{\Delta t} = -\frac{(p+1)\, x_{i-1}^p}{x_i^{p+1} - x_{i-1}^{p+1}} \frac{\theta_i - \theta_{i-1}}{\Delta x} + \frac{(p+1)\, x_i^p}{x_i^{p+1} - x_{i-1}^{p+1}} \frac{\theta_{i+1} - \theta_i}{\Delta x} \qquad (4)$$

Here the numerical value of p is as defined earlier and **does,** where shown as a superscript, indicate exponentiation. Use of the general form shown in Equation 4 means that a single algorithm may be deployed for all three geometries. Special forms of this equation used for the node at the centerline (where the first term on the right drops out) and that at the surface (where the second term on the right is modified to represent convection to the fluid) may be found in the on-line help files. The surface treatment does introduce another non-dimensional parameter, the Grid Biot number:

$$Grid\ Bi = \frac{h\,\Delta x}{k} \qquad (5)$$

By our choice of grid spacing internal to the module ($\Delta x = 1/20$) this Grid Biot number (or finite-difference form of the Biot number) is always 1/20 of the standard Biot number (Eqn. 3) and thus is <u>not</u> a concern of the user.

3.1 One-Dimensional, Transient Conduction

The *explicit* representation as expressed in Equation 4 will produce non-physical numerical oscillations if too large a timestep is used for a given spatial increment. For this reason the program was written in a *semi-implicit* form, allowing the user to select a strictly explicit or *implicit* formulation or some weighted combination of the two. The explicit timestep restriction for the infinite wall is written for internal nodes as:

$$\text{Grid Fo} = \frac{\alpha \Delta t}{(\Delta x)^2} \leq 0.50 \qquad (6)$$

Similar restrictions are computed internally for the other geometries. The Grid Fourier number is the non-dimensional time increment and is a user input.

Program Operation

This module was developed using a seamless integration of the Visual Basic and Fortran programming languages. Watcom Fortran 77, which includes many Fortran 90 extensions, is used for the intensive numerical computations and for generation of the transient display. The applicable Fortran routines are combined into a single dynamic link library (DLL) that is then available for function-calls from a tailored executable written in Microsoft Visual Basic. A typical view of the interface (at the completion of a transient calculation) is shown in Figure 2.

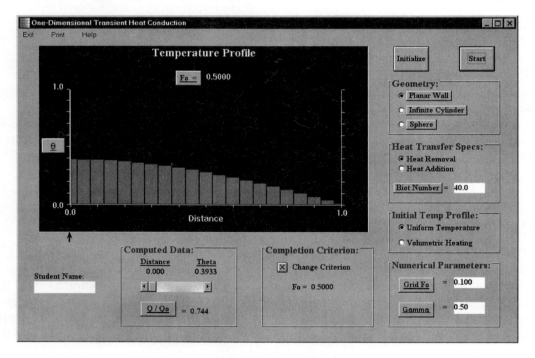

Figure 2. Interface after a calculation

All user input and output are through the Visual Basic interface. In addition, several "Hot Buttons" are available to provide a quick definition of the various input parameters (Biot number, Grid Fourier number, etc.) and outputs (the temperature θ, total heat transferred, etc.) More thorough explanations are given in the on-line help files. In the following text bolded items refer to controls and inputs on the user interface.

3.1 One-Dimensional, Transient Conduction

Geometry

In the top panel of the display (Figure 2), the user can select one of three possible geometries:

- **Plane Wall** (infinite slab),
- **Infinite Cylinder**,
- **Sphere**.

The transient heat-transfer process is perceptively different for each of the three geometries, as will be noticed when viewing the dynamic changes in the relative temperature profiles.

Heat Transfer Specifications

In the second input panel from the top, the user must specify the direction of the heat exchange between the solid material and the contacting fluid. **Heat Removal** implies that the material begins at a temperature hotter than the fluid, while **Heat Addition** is, of course, used to specify the reverse process. Just as the Heisler Charts are used interchangeably for both heating and cooling, there is no difference in the calculation for these two cases, only in the transient display. This panel also provides for the manual input of the surface **Biot Number**, the non-dimensional parameter relating the internal conductive resistance of the solid to the thermal resistance of the convective boundary-layer (See Equation 3). For a very large Biot Number (small convective resistance, which corresponds to large h), the dynamic plot of temperatures will show an immediate step-change at the surface node to match the fluid temperature.

Initial Temperature Profile

Each problem solution may be initialized with either of two available temperature distributions:

(a) **Uniform Temperature** throughout the solid material;

(b) **Volumetric Heating** wherein the material is subjected to uniform steady-state volumetric heating that is suddenly turned off at initialization of the transient computations. The same Biot number is used in determining the steady-state temperature distribution as in computing the transient profiles. This volumetric heating initial profile is only available in the **Heat Removal** mode.

Clicking on the **Initialize** button will generate a bar chart of the selected profile.

Numerical Parameters

While all other parameters (Biot number, Fourier number) are identical to those one would compute in using the Heisler charts, the module being a numerical solution in addition requires the user to provide two parameters associated with the numerical computations themselves.

3.1 One-Dimensional, Transient Conduction

The **Grid Fourier Number** is a non-dimensional time step based on increments of actual time and the spatial increments defined by the grid nodal arrangement. When fully explicit computations are specified, this number must be chosen with care in order to avoid numerical instability. With the grid spacing (Δx) of 1/20 "hardwired" into the program, a single timestep (Δt) translates to a change of 1/400 that number in the overall Fourier number. For instance, if one were to specify a grid Fourier number of 0.5, then it would take 800 timesteps to reach an overall Fourier number (elapsed time) of 1.0.

The **Gamma** parameter allows the user to specify a weighted combination of explicit and implicit discretization. (See on-line help files for details.) With a fully explicit solution ($\gamma = 0$), the user must beware of the numerical instability mentioned above. This difficulty may be avoided by specifying some combination with the fully implicit ($\gamma = 1$) algorithm. For any degree of implicitness (i.e., $\gamma > 0$), a tridiagonal system of linear equations is automatically set up and solved at each time step. With a fully implicit solution, any time step will be stable, so that it is up to the user to specify a Grid Fourier Number small enough to yield the desired accuracy. A value of $\gamma = 0.5$, which gives equal weighting of old and new values in the representation of the conductive fluxes, will generally be the best choice. This equal weighting is known as the Crank-Nicolson method.

Completion Criteria

After clicking on the **Start** button, the program runs until either of the two possible stopping criteria is met:

(a) The specified **Fourier Number** is reached. This number represents non-dimensional elapsed time and directly controls the interval during which the computations will continue.

(b) The specified non-dimensional temperature, θ, is reached at a chosen location in the material. Upon completion, the corresponding Fourier Number (elapsed time) for the computation is displayed along with the plot of temperature profile.

Clicking on the **Change Criterion** button in the bottom center of the display opens another form on which the user specifies these criteria.

Temperature Profile

After clicking on the **Begin** button, a dynamic display of the temperature profile, as a function of time and position, is provided in the large plotting window. The length of each bar is altered after each time step to reflect the progress of the transient. All variables used in the program are non-dimensionalized. Thus, the centerline of the infinite wall, cylinder or sphere is to the left of the plot, while the material surface exposed to convection is to the right at Distance = 1. The non-dimensional temperature, θ, is based on values of the initial material temperatures and the temperature of the fluid in contact with the surface.

Note that for problems in which the material is being heated by the fluid, the ordinate of the profile plot is re-defined as $(1 - \theta)$. In this manner, the length of the vertical bars are an indicator of the heat remaining in the solid during a cooling process and of the heat that has been added to the solid during a transient heating process.

At the conclusion of a numerical solution, the final numerical values of θ are available at all locations using the scroll bar in the **Computed Data** panel located below the plot window. (Heisler chart users obtain this data from a combination of the first and second chart of the series.)

Total Energy Transfer

Also at the end of the calculation, the **Computed Data** panel displays the ratio of total heat transferred relative to the maximum energy that could be transferred were the process to be continued indefinitely. This ratio is often included as the third plot of the series [2, 6]. At least for the infinite wall, where there is a linear relationship between the "amount of bars" left on the screen at the end and the energy still stored versus that at the start, the user can immediately sense the physical meaning of this parameter.

Program Verification

As with all software, some means of assessing the validity of the computed solutions must be employed. Fortunately with analytical solutions readily available for all three geometries, as well as for the more restricted problems discussed below (lumped capacitance, semi-infinite solutions, etc.), the verification task for this module is relatively straightforward. Comparison of at least some computed results with those obtained from the standard means should be a vital part of any projected use of this program.

For the uniform initial-temperature option, the module solves exactly the same problems as covered by the Heisler/Gröber charts. Comparison with the charts therefore provides a relatively easy verification of the program's numerical results. In fact, the numerical solution presented here should be more accurate than the charts, which do not even apply for values of Fourier number less than about 0.2 because the charts are based on only a single-term approximation for an infinite-series solution.

Another check, which was implemented into the program code, uses a computation of an integral heat balance. If one takes the initial heat content in the material and subtracts off the summation of the heat transferred out the surface over the period of the transient, the difference ought to match the energy content remaining in the material volume. This check was used in the development of the module and results match to at least five significant figures. Such a balance is a necessary, but not sufficient condition.

References

1. R.J. Ribando and G.W. O'Leary, "A Teaching Module for One-Dimensional, Transient Conduction," *Computer Applications in Engineering Education*, Vol. 6, pp. 41-51, 1998.

2. F. P. Incropera and D. P. DeWitt, *Fundamentals of Heat and Mass Transfer*, 4th Ed., Wiley, New York, 1996.

3. A. F. Mills, *Heat Transfer*, Irwin, Homewood, Ill, 1992.

4. M. P. Heisler, "Temperature Charts for Induction and Constant Temperature Heating," *Transactions of the ASME*, Vol. 69, 227-236, 1947.

3.1 One-Dimensional, Transient Conduction

5. R. J. Ribando and G.W. O'Leary, "Teaching Modules for Heat Transfer," *ASME Proceedings of the 32nd National Heat Transfer Conference," Vol. 6, Innovations in Heat Transfer Education and Student Heat Transfer Designs,* Edited by M.V.A. Bianchi, P.M. Norris, A.M. Anderson and A. Duncan, ASME, New York (1997).

6. H. Gröber, S. Erk and U. Grigull, Fundamentals of Heat Transfer, McGraw-Hill, New York, 1961.

7. K. W. Childs, "HEATING - A Multidimensional Finite-Difference Heat Conduction Code," http://www.cad.ornl.gov/cad_ce/text/kch1.html.

3.1 One-Dimensional, Transient Conduction

Appendix: Some Additional Insights and Applications

In addition to all problems normally solved using the Heisler Charts, the module has value for a number of other heat transfer situations. Indeed, along with the extensive help files (Figure 3), it may be used for nearly all problems typically addressed in the chapter on transient conduction found in all heat transfer texts. Multi-dimensional, transient problems that involve property variations, etc. may be approximately solved using the *Heating* program [7] as well as commercial codes.

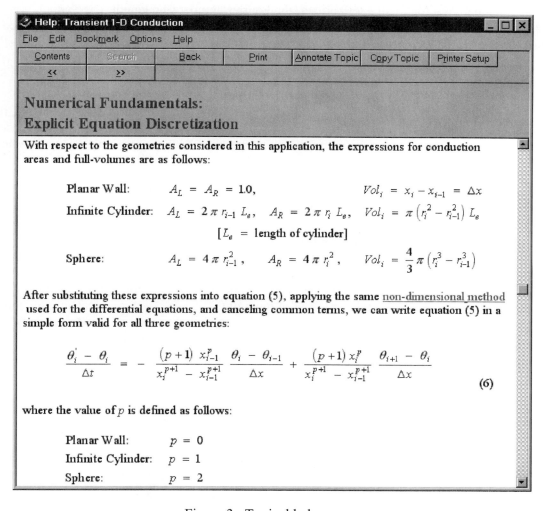

Figure 3. Typical help screen

Lumped Capacitance Assumption

Most heat transfer texts (and many differential-equations books) describe the *lumped-capacitance* or lumped-mass method, which is applicable to those cases in which the temperature gradients within the solid are small, i.e., when nearly all the resistance to heat transfer is within the surface convective-film rather than in the solid itself. In these cases an ordinary (as distinct from a partial) differential-equation applies to the transient-conduction process. Typically, values of the Biot Number less than about 0.1 are sufficient to result in negligible error when using this approximation in lieu of the partial differential-equation.

3.1 One-Dimensional, Transient Conduction

The user can input small values of Biot Number to the module and immediately see the rationale for this approximation. The user should note that for conditions under which the lumped-capacitance approximation is valid, the transient temperature is given (see Reference [2], Equation 5.13) by the expression:

$$\theta = e^{-Bi\ Fo}$$

Here *Fo* is the Fourier number (the non-dimensional lapsed time). In order to effect a significant change in temperature for small Biot number, the Fourier number (the length of the simulation run-time) must therefore be correspondingly large. For such long simulations, the user is urged to increase the Grid *Fo* (non-dimensional timestep) and to increase the value of γ to allow a more implicit computation. As noted earlier -- with our grid spacing of 1/20 -- a Grid *Fo* of 1.0 will necessitate 400 computational time-steps to achieve a unit change in Fourier number.

While this module was designed only for infinite slabs, infinite cylinders and spheres, the lumped-capacitance method remains useful for less regular geometries, as well as for those cases where, for instance, a radiative (non-linear) or time-dependent boundary condition is applicable at the surface.

Semi-Infinite Solutions

Mills [3] suggests that for Biot numbers less than 0.1 and Fourier numbers less than about .05, the semi-infinite solid solution is applicable. If we use the criterion that our finite-domain calculation is invalid for a semi-infinite problem when the temperature at the adiabatic surface (x = 0) has changed by more than, say 1%, (and use that value as the stopping criterion), we find a stopping Fourier number very close to his value for cases of large (100+) Biot number. The number increases to about .250 for a Biot number of 0.1. (With the numerical solution there is no problem associated with Fourier numbers less than 0.2.) In any case the module clearly demonstrates the applicability and rationale behind the semi-infinite solutions.

3.2 Monticello Problem

Introduction

Even though their "R" value is quite low, the thick masonry walls of an old structure like Monticello have an enormous effect on the heating and cooling load of the building. Mr. Jefferson and his contemporaries in the early nineteenth century didn't understand the insulating value of air immobilized in fibrous materials, a phenomenon we commonly exploit today in the form of fiberglass and other lightweight insulations. But they, and earlier civilizations all over the world, certainly knew about and used the effects of thermal mass. In more recent years manufacturers of fiberglass insulation have compared the "R" value of their product with that of brick and masonry and have claimed correctly that it is vastly superior. The masonry industry has countered (again correctly) that in assessing the value of its products, you must consider transient behavior and that the "R" value, a steady-state property, is therefore, not the appropriate measure. We will study the matter here.

Figure 1. Monticello, Photo courtesy J.H. Arthur

At Monticello the walls are made of 16" of solid brick covered directly with 1" of plaster on the inside. You can see this same construction "au naturel" at Mr. Jefferson's retreat, Poplar Forest, which is currently under restoration outside Lynchburg, Virginia. This is certainly not a construction you would find in most contemporary residences, but some modern homes do incorporate energy storage in the form of poured concrete or rocks. In this problem we will study the effect of this large thermal mass by numerically solving a simple 1-D, transient conduction problem with appropriate property values and boundary conditions.

3.2 Monticello Problem

This problem is a relatively simple example of how the textbook charts based on analytical solutions to the transient, one-dimensional conduction equation are often not applicable. While the analytical solutions are only appropriate for a sudden, step change in thermal boundary conditions, in this case one of the boundary conditions we wish to apply is a periodic function of time. Also, the two sides of the slab are exposed to different convection conditions, a scenario that again is not covered by the standard "chart" solutions.

Implementation

In this project we will develop a 1-D, finite-difference (or finite-volume) model to study the transient storage effects in Monticello's walls. For the wall itself we will use a grid spacing of .0432m (1.7"), giving nine full-sized nodes plus half-sized nodes at each end (See sketch). That setup means 11 unknown temperatures and three separate predictive equations that you will have to derive. We will use the thermal properties of face brick (the ones without the holes) for the whole thickness. (You will notice in the property tables that face brick has a higher density and higher thermal conductivity because it lacks the dead air space in the holes. You can use the specific heat for common brick.)

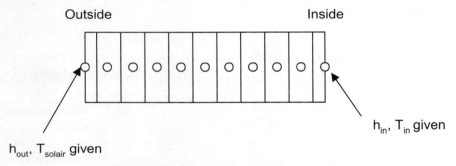

Figure 2. Nodalization of Wall

At the inside surface let us assume $h = 8$ W/m^2 K with a constant air temperature of 21°C. (We are assuming that Mr. Jefferson keeps his air conditioning set at this value.) For the outside we must include both radiative and convective gains and losses. Rather than treat these gains and losses separately, it is common (and vastly easier) to use an effective *Sol-Air* temperature, which, when put into the standard Newton's law of cooling ($q = h\,A\,\Delta T$), has the equivalent effect. The Sol-Air temperature is a function of the date, hour, latitude, and the radiative properties and orientation of the wall with respect to the sun. We will use values for a dark, west-facing wall at 40° N latitude on July 21. Hour-by-hour values for these conditions as given in the ASHRAE Handbook of Fundamentals are as follows:

24.4, 24.4, 23.8, 23.3, 23.3, 25.0, 27.7, 30.0, 32.7, 35.0, 37.7, 40.0,
53.3, 64.4, 72.7, 75.5, 72.2, 58.8, 30.5, 29.4, 28.3, 27.2, 26.1, 25.0

Here the first value (24.4°C) corresponds to 1:00 a.m.; the last value (25.0°C) corresponds to midnight. Note that the effective Sol-Air temperature gets up to nearly 170° F (75.5°C) in the afternoon. (We are assuming there are no trees shading this wall.) If you wind up using a timestep of less than an hour, it would be advisable to fill in the intermediate values using linear interpolation. Otherwise the heat flux vs. time plot you will eventually plot will exhibit some non-physical "kinks" in response to the sudden step change in boundary condition every hour.

3.2 Monticello Problem

Use a convection coefficient h = 18 W/ m² K at the outside surface. (The higher outdoor value reflects the effect of breezes that are presumably not present indoors.) You can start your simulation from arbitrary initial conditions, but make a reasonable guess. Because you don't know the actual temperature distribution at the time you begin your simulation, you will have to run it through several days to get to a steady, periodic condition. (Hint, determine the characteristic conduction time of the wall (L^2/α) to give you an estimate of that time. This estimate can be found by remembering that the time constant for an electrical circuit having both resistance and capacitance is given by t = RC. Here the thermal resistance is R = L/k and the thermal capacitance is $C = L\rho c_p$ and, of course, they are distributed and commingled, not lumped.) For Mr. Jefferson's wall L = .432 m (17") and you can compute α, the thermal diffusivity, from the thermal conductivity (k), density (ρ) and specific heat (c_p) values you find for brick. (We're assuming the 1" plaster layer inside has thermal properties similar to those of brick.) Then plot for one full day the heat transfer at both the inner and outer surface as a function of the time of day. This means that in order to make the plot, you will have to save at least the predicted temperature of the outside surface and that of the inside surface for each hour of the last day of your simulation. The heat transferred at the inner wall divided by the overall heat transfer coefficient for the wall will give you the hourly "cooling load temperature difference" (CLTD) commonly used in load estimating.

In this problem you have convective boundary conditions at both sides and a time-dependent T_∞ at the outside. If you were doing this project in a high-level programming language, you would need an outer repetition structure ("Do Loop") covering the number of days you need to run. The next Do Loop in will cover the 24 hours of the day. Whether or not there is yet another one inside this depends on whether you decide to implement an *explicit* or *implicit* solution. When you compute the allowable explicit timestep for the given material properties and grid spacing – and you do need to check it both for internal cells and for the surface cells – you will find that it is less than one hour. If you do decide to implement an explicit solution, then pick a timestep that gives an integral number of steps in each hour and is stable. Inside this will be another Do Loop to cover the nine internal nodes, plus the special equations needed at each end. All heat transfer texts include derivations of the explicit forms of the appropriate predictive equations for both internal cells and for the half-cells at the two surfaces exposed to convection and discuss the timestep limitation. Your report should include the derivation of these equations.

If you want to see the enormous effect of the direct solar gain, you may want to try using Sol-Air data for a dark, north wall in July under the same conditions as a comparison:
24.4, 24.4, 23.9, 23.3, 23.3, 32.3, 32.2, 30.6, 32.8, 35.0, 37.8, 39.4,
41.1, 41.1, 41.1, 40.0, 42.2, 41.7, 30.6, 29.4, 28.3, 27.2, 26.1, 25.0

Reference

1997 ASHRAE Handbook - Fundamentals, American Society of Heating, Refrigerating and Air-Conditioning Engineers, Inc., Atlanta (1997).

montiex – 5/26/00

3.2 Monticello Problem

Appendix I - Some Specific Hints

Fortran with *Explicit* Differencing

If you want to program this assignment in a high-level programming language and use an explicit approach, then this section is for you. You can probably implement this assignment in 30 or fewer lines of executable code. The loop structure for your program has already been discussed. There is no particular reason to save the huge amount of data your program will generate; so you really only need two vectors, each of them 11 entries long to hold the temperatures of the 11 cells. At each timestep you will calculate the new temperatures in terms of the old ones. Once you have completed all 11 cells, those new ones can be transferred into the vector of old values, the time incremented and you are ready to continue marching out in time. You will need to save the appropriate data to make the plot at the end.

Spreadsheet with *Explicit* Differencing

This problem is very easy to set up on a spreadsheet. You would probably set up 11 columns for the nodal temperatures plus a column for the T_∞ specified earlier for the outside, and one for the (constant) indoor temperature. Each row (and there will be a whole lot of them!) will represent a different time. Essentially, you will need a row for each timestep, and that will mean several days times 24 hours per day times the number of time steps you need per hour. At the top of your spreadsheet, you probably will want to input certain parameters, e.g., the grid Fourier number, the grid Biot number at the outside and that at the inside. When you use those input parameters in the cell formulae, don't forget to use absolute, as opposed to relative, addresses, or better yet, "name" the cells containing these parameters. You will be duplicating the predictive formulae many, many times. This approach isn't very elegant, but it works, and it is easy to generate the necessary plots at the end.

As a first step you may want to just set the temperature of Cell 1 to the current "Sol-air" temperature and that of Cell 11 to the interior temperature. Then once you are sure you have the formulae for the interior cells (2-10) correct, you can implement the special forms you have derived for the surface cells.

Fortran with *Implicit* or *Semi-implicit* Differencing

Most texts cover the use of an implicit differencing scheme for 1-D problems like these. In an implicit scheme the terms representing fluxes in and out by conduction or convection are expressed in terms of temperatures at the new time level rather than the old one. The predictive equations involve new (unknown) values of temperature at the node itself and each of its next-door neighbors. With an implicit scheme, you can use whatever timestep you like, so the desire for accuracy, rather than numerical stability limitations, dictates your choice. Here, where you are given the Sol-Air temperature only every hour anyway, that would be a reasonable timestep, rather than the fraction of an hour found necessary earlier for an explicit calculation. Better even than an explicit or implicit solution is a *semi-implicit* procedure in which the fluxes are expressed in terms of a weighted average of temperatures at the old and new timesteps. In particular the Crank-Nicolson method uses equal weighting of old and new and is easy to implement.

3.2 Monticello Problem

The downside of using an implicit or semi-implicit approach for this 1-D problem is that at every timestep you have a tridiagonal system of equations to solve. While you may have implemented the Thomas algorithm for tridiagonal systems as an exercise in a programming class at some time or other, it is wise to stick to well-documented, debugged and optimized software such as the SGTSV subroutine from the LAPACK collection. You can download this subroutine at http://www.netlib.org/lapack/index.html. Alternatively, you can "solve" the tridiagonal system by iteration. To use iteration, you would solve the predictive equations for the point in terms of its neighbors to the left and right and its own value at the previous timestep, then sweep over the set of equations several times, gradually improving the unknown values each pass through. The more equations you have (here you have 11 to iterate at each time step), the more sweeps will be needed, so you could wind up using as much CPU time as you would have with an explicit scheme and smaller timesteps.

Spreadsheet with *Implicit* or *Semi-implicit* Differencing

You can implement the implicit or semi-implicit scheme on a spreadsheet also. Again you would probably pick a timestep of an hour, rather than the fraction of an hour needed if you use an explicit method. Input the governing equations (internal and special forms at the ends) for the unknown temperatures. Each will involve its two unknown neighboring temperatures plus old (known) values at itself and (if you are using a semi-implicit formulation) the same two neighbors. (We say this involves a 6-point *stencil*.) Your spreadsheet will balk at these forms since they involve unknown values - and maybe put up a flag that says "circular reference," but all spreadsheets have a recalculation or iteration option to handle situations like this. (This option can be used to for elliptic (boundary-value) equations - like those that describe 2-D steady-state conduction.)

Some spreadsheets allow the user to specify the desired sweep direction for the iteration – across, up and down, or both. If yours does have that option, you should sweep across only. You are solving a "parabolic" partial differential equation here. Time is a "one-way" coordinate and once you have gotten a solution at a particular point in time, there is no justification for repeatedly recalculating it.

3.2 Monticello Problem

Appendix II – Periodic Conduction in a Semi-infinite Body

You can gain some additional physical feeling for the effects of thermal mass by running the Excel spreadsheet for Periodic Conduction in a Semi-infinite body available on the CD. A "semi-infinite" body is one thick enough that any change in the thermal boundary condition

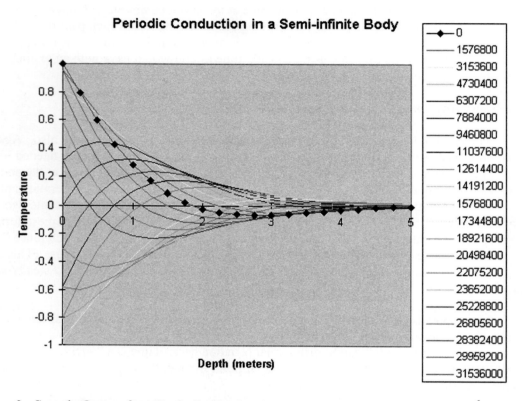

Figure 3. Sample Output from Periodic Conduction Spreadsheet - One year (31.5×10^6 seconds) is covered by the calculation. Thermal properties typical of soils are used

implemented on one side never penetrates to the other boundary. This workbook, an implementation of the analytical solution [2], allows you to change the thermal properties of the material, the period of the sinusoidal surface temperature variation, etc. Knowing the amplitude of the annual surface temperature variation and the thermal properties of the soil, one could in theory determine how deeply a water pipe should be buried to lie safely below the frost line.

Reference

Gebhart, B., *Heat Transfer*, McGraw-Hill, New York (1971).

3.3 Sandwich Wall Construction

Introduction

Figure 1. Sandwich wall under construction

Recently an alumnus designing his own dream home decided to use what he had learned about thermal storage from having done the previous Monticello project as a student. The cost of the materials and labor involved in building 16" thick brick walls like those at Monticello would, of course, be prohibitive today. On the other hand, since Jefferson's time we have recognized the insulating value of air immobilized in fiberglass and other lightweight insulations. This new home, shown under construction in the photo, uses a construction that combines the advantages of concrete as a thermal storage medium with lightweight Styrofoam as an insulating material.

In this construction rigid Styrofoam boards are used as the forms when the concrete is poured, but unlike normal plywood forms, which are removed for reuse elsewhere when the concrete has hardened, the Styrofoam is left there permanently. The final construction thus consists of two layers of lightweight Styrofoam, each having a significant insulating ("R") value, plus a layer of concrete having a large thermal mass in the space between them. In the basement the Styrofoam layers are 2.375" thick and the concrete is 7.875" thick, while in the living space 6.25" of concrete is sandwiched between 2.625" thick sheets of Styrofoam. A total of 136 cubic yards of concrete were used in the 4400 sq. ft. (finished basement and main level) home. Standard surface treatments are used inside and out.

In our formulation of this problem, we will ignore the steel "rebar" that is seen protruding from the concrete in the photo. The braces seen above on the inside are shipped back to the supplier of the kit once the concrete work is complete. The black vertical stripes seen on the Styrofoam are nailing (furring) strips and are on the outside as well, allowing the buyer to finish both the inside and outside surfaces as they so choose. Other than the deeper than usual window and door penetrations, the finished construction looks quite conventional.

The foam-concrete-foam configuration seen in the photograph above lends itself readily to a simple one-dimensional analysis; other designs use the same principle of combining the thermal capacity of concrete with the insulating value of Styrofoam. The analysis of a system such as that shown to the right as one-dimensional would, however, be problematic.

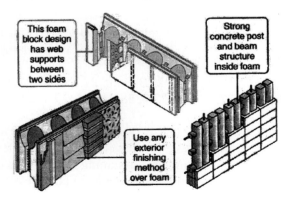

Figure 2. Alternate foam/concrete construction [1]

3.3 Sandwich Wall Construction

Another contemporary construction, autoclaved aerated concrete, combines the insulating value of trapped air more directly with the heat capacity of masonry [2,3]. These blocks are pre-cast at the factory, where a chemical reaction that takes place before the concrete has set creates thousands of tiny bubbles throughout the material. These lightweight blocks have been used elsewhere, particularly in Europe, for over half a century and are now becoming more popular in residential and commercial construction in the southern U.S.

Implementation

In this project we will develop a 1-D, finite-difference model to study the transient storage and heat transmission effects in the Styrofoam/concrete composite wall. We will use a grid spacing of 0.5″ (.0127 m) and approximate thickness values in the numerical implementation in order to simplify the nodalization. Let us assume that each piece of Styrofoam is 2 ¾″ thick so that we have 5 ½ nodes in the insulation. Assuming that the concrete is 6 ½ ″ thick, then we have 13 full cells in the middle layer. The entire computing region will then have 23 full cells, plus half-cells at each end, giving a total of 25 unknown temperatures (See sketch).

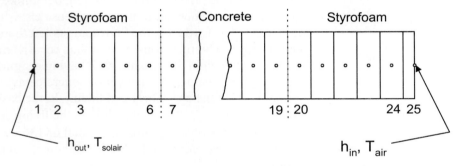

Figure 3. Nodalization for finite-difference calculation

We will treat the surface boundary conditions the same way as in the earlier Monticello problem. At the inside surface let us assume h_{in} = 8 W/m² K with a constant air temperature of 21°C. For the outside we must include both radiative and convective gains and losses. Rather than treat these gains and losses separately, it is common (and vastly easier) to use an effective temperature, which when put into the standard Newton's law of cooling ($q = h\,A\,\Delta T$) has the equivalent effect. The Sol-Air temperature is a function of the date, hour, latitude, and the radiative properties and orientation of the wall with respect to the sun. We will use values for a dark, west-facing wall at 40° N latitude on July 21. Hour-by-hour T_{solair} values for these conditions as given in the ASHRAE Handbook of Fundamentals [4] are as follows:

24.4, 24.4, 23.8, 23.3, 23.3, 25.0, 27.7, 30.0, 32.7, 35.0, 37.7, 40.0,
53.3, 64.4, 72.7, 75.5, 72.2, 58.8, 30.5, 29.4, 28.3, 27.2, 26.1, 25.0

Here the first value (24.4°C) corresponds to 1:00 a.m.; the last value (25.0°C) corresponds to midnight. Note that the effective Sol-Air temperature gets up to nearly 170° F (75.5°C) in the afternoon. (We are assuming there are no trees shading this wall.) If you wind up using a timestep less than an hour, it would be advisable to fill in the intermediate values using linear interpolation. Otherwise the heat flux vs. time plot you will eventually obtain will exhibit some non-physical kinks in response to the step change in boundary condition every hour. Use a convection coefficient of h_{out} = 18 W/ m² K at the outside surface. (The higher outdoor value reflects the effect of breezes that are presumably not present indoors.)

3.3 Sandwich Wall Construction

At the end you will want to plot the heat transfer at both the inner and outer surfaces of the wall. This means that in order to make the plot, you will need the predicted temperature of the outside surface and that of the inside surface for each hour of the last day of your simulation. The heat transferred at the inner wall divided by the overall heat transfer coefficient for the wall will give you the hourly "cooling load temperature difference" (CLTD) commonly used in load estimating.

Numerical Formulation

We will set this problem up using an *implicit* formulation for several reasons: (1) we will want to run this simulation over a fairly long period of time, (2) we have the two internal material interfaces to handle, and (3) we have the convective boundary conditions at both ends. With the implicit solution we will be able to use relatively long time steps, in fact, a full hour will be very convenient.

Let us consider a generic internal cell first. For such a cell the discretized heat balance equation written in an implicit form reads:

$$\left(\rho c_p\right)_i \frac{T_i' - T_i}{\Delta t} \Delta x = -k_l \frac{T_i' - T_{i-1}'}{\Delta x} - (-)k_r \frac{T_{i+1}' - T_i'}{\Delta x} \tag{1}$$

Here the primes indicate the advanced (new) time value. Normally we would divide through by the density-specific heat product, assume k is uniform and introduce a grid Fourier number,

$$Fo = \frac{k}{\rho c_p} \frac{\Delta t}{(\Delta x)^2} = \alpha \frac{\Delta t}{(\Delta x)^2}. \tag{2}$$

Here, although the density-specific heat product for each cell is uniform, we need to be able to handle different thermal conductivities for the inward flux at the left and the outward flux to the right (specifically for nodes 6,7,19 and 20 which are adjacent to the material interfaces). That being the case, here instead we divide through and introduce the coefficients Cl_i and Cr_i (which for most cells <u>are</u> simply the grid Fourier number):

$$T_i' = T_i + Cl_i\left(T_{i-1}' - T_i'\right) + Cr_i\left(T_{i+1}' - T_i'\right) \tag{3}$$

Now solve for T_i':

$$T_i' = \left[T_i + Cl_i T_{i-1}' + CR_i T_{i+1}'\right] / \left[1 + Cl_i + Cr_i\right] \tag{4}$$

Thinking now in the context of a spreadsheet with the spatial coordinate (x) running across and time increasing downward, we have a cell formula relating the temperature at a point with the (known) temperature above it (T_i) and the unknown temperatures to its left (T_{i-1}') and right (T_{i+1}').

3.3 Sandwich Wall Construction

When one derives the appropriate form of the heat balance equation for the half cell at the left exposed to convection (Cell #1), one realizes that it can also be put in the form of Equation 3, but with T'_{i-1} replaced with the known T_{solair}. The same principle applies at the interior surface. Thus, the same cell formula may be used for all cells. Most work is involved in determining the proper coefficients Cl and Cr for all 25 cells, including the generic internal cells, the ends, and interfaces. These coefficients may be pre-computed and stored in two rows at the head of each of the 25 temperature columns. Don't forget how you must address these values in your cell formulae – one direction will have a relative address and the other an absolute address.

During the early development of your spreadsheet you might want to concentrate your efforts on the internal cells and, only once they appear to be computed correctly, moving on to the surface cells. Temporarily setting the outside surface cells to $T_{solair}(t)$ and the inside surface temperature to the inside air temperature makes a good strategy.

Additional Spreadsheet Notes

With cell formulae like that in Equation 4, your spreadsheet will give an error message informing you that you have a "circular reference." Indeed, what you are really trying to solve here is a *tridiagonal* system of linear equations at each timestep (Δt). If you were to write out the complete system of 25 equations in 25 unknown temperatures that you need to solve at each time step, you would soon realize that each unknown temperature depends only on its immediate neighbors. Written in matrix form **Ax = B,** the coefficient matrix **A** would have a bandwidth of three. You could implement a direct solution technique for a tridiagonal system based on the well-known Thomas algorithm [5] in only about 15 lines of code, but here we'll have the spreadsheet solve them approximately using iteration.

You can wait for your spreadsheet to tell you that you have a "circular reference" (meaning that you have cell formulae that depend on other unknown values) and follow its directions as to what to do about it, or you can take a proactive approach and specify that it should iterate. You can specify how many iterations to allow and how tight a tolerance to use. Don't use a huge number of iterations at first, nor an excessively small tolerance. Later, after you have gotten this project working you will want to crank up the number of iterations (into the thousands) and tighten up the tolerance (e.g., to .00001). Ideally you would solve the tridiagonal system completely at each timestep before moving on to the next, but unfortunately some spreadsheets (including Excel) don't give you any control over the direction of the iteration that it will do for you. For those that do (e.g., Quattro Pro), you will want to iterate thoroughly over each row before moving down to the next one.

When we did the original Monticello problem (17" of solid brick), we simply started the calculation from a guessed initial state and ran it through four or five days to get to a fairly steady periodic state. We copied a single day's worth of cell formulae (24 rows) and boundary conditions and pasted it several times at the bottom. With the configuration used in the current problem, it takes a fairly long time to reach a steady periodic state (physically), plus Excel does its iteration in both directions. So that approach seems pretty unwieldy. Since we are mainly interested in the steady, periodic condition, we recommend that you insert another row just above your 1:00 a.m. row. For that row you'll specify that it be equal to the value it computes for that cell at midnight (down at the bottom). Eventually after enough iteration you'll have a 24-hour period of the steady, periodic solution you're looking for – and the entire spreadsheet, including headings, will have less than about 30 rows. The downside is that you will not have a feeling for how long (in days) it takes for this construction to reach a steady, periodic condition.

3.3 Sandwich Wall Construction

One way of watching this calculation converge is to set up columns where you will compute the heat fluxes at the inner and outer surfaces. When it has finally converged, those two fluxes summed over 24 hours ought to be equal – and when you see that you will have a lot of confidence in your solution!

References

1. Dulley, James T., "Concrete/foam block construction combines stability and efficiency," *The Daily Progress,* Charlottesville, Va, Mar. 23, 2000.

2. Jardine, K. and Cameron, J., "Building for Affordability and Energy Efficiency," *Fine Homebuilding*, Spring/Summer 1999, pp. 82-87

3. Bukoski, S.C., "Autoclaved Aerated Concrete: Shaping the Evolution of Residential Construction in the United States," Construction Engineering and Management Program, Georgia Institute of Technology, August 14, 1998.

4. *1997 ASHRAE Handbook - Fundamentals*, American Society of Heating, Refrigerating and Air-Conditioning Engineers, Inc., Atlanta (1997).

5. Chapra, S.C., and Canale, R.P, *Numerical Methods for Engineers*, 2nd Ed., McGraw-Hill, New York, 1988.

Sandwich 7/14/00

3.4 Transient Conduction at the Interface between Two Materials

Introduction

In this exercise we will use the finite-difference (finite-volume) method to solve a transient, 1-D conduction problem that models the contacting of two dissimilar materials having different initial temperatures. Let us assume that one of the materials represents your finger, while the other is one of the several inorganic materials we want to test. We will start with the temperature of the other material high enough that you would expect to experience discomfort and possibly a burn if you touched it long enough. Then we will run the calculation for a brief period of time and compare results for three materials.

Implementation

A sketch of the 1-D geometry is shown below. For a start let us take the grid spacing $\Delta x = .0001$ m and place 20 nodes in the section representing your finger and 20 in the other material. (You should design your spreadsheet or program so that these numbers can be changed easily.)

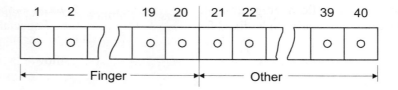

Figure 1. Nodal configuration

The appropriate density, specific heat and initial temperature may be readily assigned to each node. The assignment of the thermal conductivity is slightly complicated by the fact that, as explained later, we need the values at the cell boundaries rather than at their centers. Let us adopt the convention that "k_i" represents the conductivity at the right edge of cell "i"; then "k_{i-1}" will be the value to its left. The assignment of values is then straightforward, except at the interface between the two materials (k_{20} for the grid arrangement shown above). Here it is probably appropriate to say that you have a layer of flesh of thickness $\Delta x/2$ in series with a layer of thickness $\Delta x/2$ of the material being tested and then assign the appropriate equivalent value. We are assuming here that there is no "contact resistance" between the two materials; you could readily include a non-zero value of contact resistance as a third resistance in series at this location. (Indeed, if you were doing a real physical experiment rather than this computer simulation, you might deliberately add some extra thermal resistance between yourself and the hot surface by wetting your finger.)

With properties denoted as 1-D vectors (in fact, the density and specific heat may be combined into one variable since they always appear as a product), the transient heat balance equation may be written in an "explicit" form as:

$$(\rho c_p)_i \frac{T_i^{n+1} - T_i^n}{\Delta t} \Delta x A = -k_{i-1} A \frac{T_i^n - T_{i-1}^n}{\Delta x} - (-) k_i A \frac{T_{i+1}^n - T_i^n}{\Delta x}$$

3.4 Transient Conduction at the Interface between Two Materials

That is, the time rate of change of stored thermal energy is equal to the difference between the energy conducted in through the left side and that conducted out through the right. Here the superscript n refers to the current time level, while n+1 refers to the advanced (unknown) time level. One negative sign in each of the conduction terms on the right comes from Fourier's Law; the other one in the second term comes from our convention that a positive flux into the left side is a net addition to that cell, while a positive flux through the right side represents a loss to that cell. This equation, when solved for the unknown T_i^{n+1}, gives a "prescription" which can be used to predict temperatures in terms of known values. Note that the cross-sectional area (A) cancels out.

Upon studying the above equation and our nodalization (Figure 1), it should be evident why the heat capacity multiplying the storage term is evaluated at cell centers while the thermal conductivity that appears in the two flux terms is evaluated at the edges. By evaluating the thermal conductivity at the edges, we can be sure that a term having equal magnitude and opposite sign appears in the heat balance we apply at the next-door neighbor - thereby ensuring that our scheme conserves energy exactly. Some authors will emphasize this placement of the thermal conductivity at the edges by designating it as $k_{i-1/2}$, $k_{i+1/2}$, etc., but unfortunately computers do not like fractional subscripts.

Start with an initial flesh temperature of 30°C and an initial material temperature of 300°C. Take the outer boundaries to be fixed for all time at these values. Especially for a high thermal conductivity material like cast iron, this assumption may be faulty. If you choose to do an explicit calculation as derived above, you will need to compute the allowable timestep (Δt). This value can be determined by solving the above equation for the unknown T_i^{n+1} and then requiring that the coefficient of the central term (T_i^n) be greater than or equal to 0. (See any heat transfer text.) The result you get from this analysis reduces to the usual restriction on timestep r (= Fo) $\leq \frac{1}{2}$ for a slab of a single material. Here the limiting timestep could arise in the finger nodes (1-19), in the other material (nodes 21-40) or at either of the hybrid nodes adjacent to the interface (20 and 21). You will need to check all four of these possibilities and then apply the minimum of the four. The particular cell that limits the timestep is not necessarily the same one for all three materials. For the soapstone and tile runs the limiting timestep will be pretty reasonable; for the cast iron it will be very small, and you might be well advised to switch to an implicit formulation.

Material properties are supplied in the following table:

Material	Density $\left(\frac{kg}{m^3}\right)$	Specific Heat $\left(\frac{J}{kg \cdot K}\right)$	Thermal conductivity $\left(\frac{W}{m \cdot K}\right)$
Flesh	1000	4181	0.37
Cast Iron	7608	400	80.2
Soapstone	2793	971	2.15
Space Shuttle tile	144.2	878.6	.06

In lieu of better information, values for water are listed for the density and specific heat of flesh; the thermal conductivity is a value for skin. (Since we are going to do this experiment on the computer rather than for real, it might be more appropriate to use the values tabulated in the appendix of most textbooks for chicken meat!) For most of their approximately 8 cm.

thickness, space shuttle tiles are made out of a material with the properties given in the table. (It is almost like Styrofoam.) There is, in fact, a thin, almost egg-shell-like layer on the outside which has a much higher thermal conductivity than that given in the table. We will neglect that layer in this analysis.

Figure 2. Sample of space shuttle tile. This piece is from the windward (under) side and coated with a high emissivity coating to enhance radiative losses.

Run your calculation out to an elapsed time of 0.1 seconds, which is probably a reasonable estimate of how long it would take you to respond if your hand were being burned. Plot your results (temperature vs. position at time = 0.1 secs.) for the three materials and comment. Is the assumption that the temperatures at the left end of the "finger" and the right end of the other material do not change during this transient good for all three cases? Can you explain why it is so much more comfortable to step with your bare feet onto a rug instead of on bare tile or porcelain at the same temperature?

Verification

As long as the elapsed time is insufficient for the transient to propagate to the outer edges of the computing region we have chosen, then an analytical solution based on two semi-infinite solids placed in contact is valid. From that solution we can solve for the interface temperature in terms of the initial temperatures of the two bodies and their thermal properties [1]:

$$T_s = \frac{(k\rho c)_A^{1/2} T_{A,i} + (k\rho c)_B^{1/2} T_{B,i}}{(k\rho c)_A^{1/2} + (k\rho c)_B^{1/2}}$$

From that solution you determine the contact temperature for each of the three cases (surprisingly enough it is **not** a function of time) and compare to what your program predicts. For space shuttle tile and the soapstone, the 0.1 second duration of the transient as suggested earlier is

3.4 Transient Conduction at the Interface between Two Materials

sufficiently short that good agreement between this analytical prediction and the numerical solution will be found. Unless you make your cast iron layer thicker than suggested, you should find that the semi-infinite approximation is <u>not</u> valid at 0.1 seconds into the transient.

Reference

Incropera, F.P. and DeWitt, D.P., *Fundamentals of Heat and Mass Transfer*, 4th Ed., Wiley, NY, 1996.

touch 7/15/00

4.1 Forced Convection on a Flat Plate

Introduction

This module stems from a student project in a graduate-level heat transfer course [1] and has been enhanced considerably over the years to the point where it might now be considered a virtual laboratory. The effects of Reynolds and Prandtl numbers on the development of the hydrodynamic and thermal boundary layers can be seen clearly in a matter of seconds. Various combinations of thermal boundary conditions can be implemented easily. Users can "witness" the transition to turbulence and "measure" the effect of the various inputs on the heat transfer coefficient. Data can be taken and correlations developed - much as if the user had access to a heat transfer laboratory better equipped than any in the world – a facility where one can switch in seconds from testing viscous oils to highly reactive liquid metals!

We begin with the continuity, horizontal (x) momentum and energy equations for an incompressible fluid in *boundary layer* form and allowing for variable transport properties [2]:

$$\frac{\partial u}{\partial x} + \frac{\partial v}{\partial y} = 0 \tag{1}$$

$$u\frac{\partial u}{\partial x} + v\frac{\partial u}{\partial y} = u_e \frac{du_e}{dx} + \frac{\partial}{\partial y}(v + \varepsilon_m)\frac{\partial u}{\partial y}, \tag{2}$$

$$u\frac{\partial T}{\partial x} + v\frac{\partial T}{\partial y} = \frac{\partial}{\partial y}(\alpha + \varepsilon_t)\frac{\partial T}{\partial y}. \tag{3}$$

Here we have used Bernoulli's equation to replace the pressure term in the x momentum equation (first term on right hand side of Equation 2). The variable u_e represents the free stream velocity and could be a function of x if, for instance, there were a free stream pressure gradient. That capability is not exercised here; for all cases $u_e = U_\infty$, the freestream velocity, and is taken as a constant. Similarly, the fluid transport properties, i.e., thermal diffusivity ($\alpha = \frac{k}{\rho c_p}$, i.e., thermal conductivity divided by density times specific heat) and kinematic viscosity ($v = \frac{\mu}{\rho}$, i.e., viscosity divided by density) are taken as constants. The quantities ε_m and ε_t are the *eddy diffusivities* of momentum and heat, respectively; both are considered to be zero for laminar flows. For turbulent flows the dependent variables including u and v, the two velocity components, and temperature (T) are all understood to be time-averaged values and the eddy diffusivities will be modeled. Buoyancy effects are not considered in this model.

Normally at this point, at least for laminar flows, one would introduce *similarity* variables, thereby converting Equation 2 from a partial differential equation (PDE) into the well-known ordinary differential equation (ODE) known as the Blasius equation. Similar transformations would convert the energy transport equation (3) to an ODE as well. However, modern numerical techniques as applied in this module, allow us to solve these equations as PDE's, thereby avoiding the various restrictions, e.g., on boundary conditions, imposed by the requirements for *similarity*.

To improve accuracy over what one would expect solving the set of PDE's on a Cartesian grid, we used the scheme suggested in Anderson, et al. [3] and in effect created a mesh

4.1 Forced Convection on a Flat Plate

which grows along with the boundary layer. A convenient transformation is based on the Blasius similarity solution for a laminar boundary layer, i.e.,

$$\eta = \frac{y}{x}\left[\frac{u_e x}{\nu}\right]^{1/2}, x' = x/L . \quad (4)$$

Here the quantity $\frac{u_e x}{\nu}$ is the local Reynolds number. The distance x is measured from the beginning of the plate (the left end in all subsequent examples). Details, including the transformed equations, may be found in References 1, 3 and 4.

The transformed versions of Equations 1-3 must be converted from PDE's into the algebraic equations that a computer can solve. Of the several methods available for *discretizing* the transformed equivalents to the parabolic Equations (2) and (3), we have chosen to implement the Crank-Nicholson scheme. This algorithm involves solving a tridiagonal system [5] for the horizontal velocity (u) at a particular streamwise station. Then the discretized transformed equivalent to Equation 1 is marched out from the wall to the freestream to determine the vertical (v) velocity. Finally another tridiagonal system, this one for the temperature, is solved. This sequence of calculations is repeated at each station from the leading edge to the downstream end of the plate. Special forms apply at the leading edge [4].

Except for requiring some care in selecting the variation of the grid spacing, the algorithm as described above is directly applicable to the solution of turbulent and transitioning flows. One must only provide *models* for estimating the local eddy diffusivities of momentum (ε_m) and heat (ε_t). For the former we have implemented the very simplest, a van Driest mixing length model [6]. The eddy diffusivity for heat (ε_t) is related to the comparable quantity for momentum by the relationship $\varepsilon_t = \varepsilon_m / \mathrm{Pr}_t$, where the turbulent Prandtl number Pr_t is modeled using the relationships given in Cebeci and Bradshaw [7].

The transition to turbulence is based on a model given by White [8] based on the work of van Driest and Blumer [9]:

$$\mathrm{Re}_{x,tr}^{1/2} = \frac{-1.0 + (132{,}500 T^2)^{1/2}}{39.2 T^2} \quad (5)$$

In this equation only, and consistent with White's notation, the symbol T represents the freestream turbulence level expressed as a percentage. For 1% freestream turbulence, this expression yields $\mathrm{Re}_{x,tr} = 500{,}000$, the value commonly used in heat transfer texts for laminar-turbulent transition. For extremely low values of freestream turbulence, an asymptotic value of about $\mathrm{Re}_{x,tr} = 2.86 \times 10^6$ is approached. Rather than an abrupt transition, as commonly assumed in heat transfer texts, we have followed the approach suggested by Cebeci and Bradshaw [7]. Two *intermittency* factors, both of which tend to reduce the eddy diffusivities in the transition region below what they would be for a suddenly-fully-turbulent flow, are applied.

It is worth emphasizing again that Equations 2 and 3 and their transformed equivalents are mathematically *parabolic* - just like transient diffusion (conduction) problems. A numerical solution of the latter involves marching forward in time from initial data. Here the equations are parabolic in space, and we march forward (along the plate) in space. This spacewise marching was dramatically demonstrated in the earliest versions of this program (circa mid-1980's) in

4.1 Forced Convection on a Flat Plate

which plotting (and computing) would commence at the leading edge and 20 minutes later would finally reach the downstream end!

If in a transient conduction problem we wanted particularly accurate values at some chosen point in time, we would march from initial conditions to that chosen time using small timesteps, knowing, of course, that all timesteps following our particular chosen time have no effect at all. That is, no information is propagated backward in time. Here the horizontal (spatial) coordinate is "one-way." The implication is that if you want particularly high accuracy at some particular point, e.g., in order to find a local heat transfer coefficient, it is advantageous to place that point at the end of your computational region. Then all 200 grid spacings are contributing to accuracy at that single point; that is, you have used as fine a horizontal grid spacing as the program will allow.

The Interface - User Inputs

Figure 1 below shows the main interface and output window. The intensive calculations, including the graphical output shown inside the black window, are done in Watcom Fortran 77. Many Fortran 90 features, including data structures and dynamic array allocation are available in the latter product and have been used extensively in this module. Boxes seen on the interface having a white background are for user input. These include the Reynolds number for the flow, Prandtl number of the fluid, and turbulence level of the free stream. Each of these parameters includes a "hot" button, i.e., by clicking on the label, the user gets a succinct explanation of that parameter. The user can also specify a plot enlargement factor in the y direction since for other than very low Reynolds numbers, a boundary layer plotted to scale would otherwise be nearly invisible.

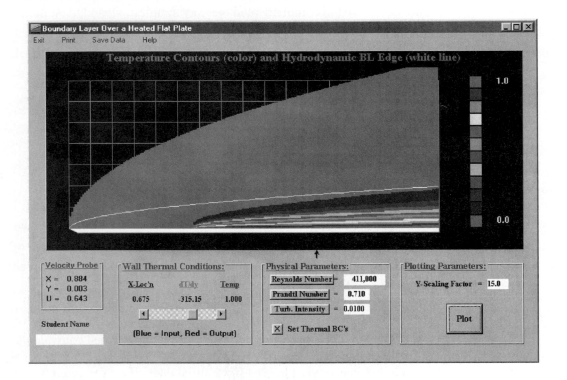

Figure 1. Interface and computed results for unheated starting length problem. Flow is from left to right.

4.1 Forced Convection on a Flat Plate

When clicked, the **Set Boundary Conditions** button opens another form where thermal boundary conditions may be specified (Figure 2). On that form the user can specify as many as five separate zones along the plate where the wall thermal boundary condition may be set. The sliders are used to set the downstream end of each zone; the upstream end of the following zone is then automatically set. For each zone the user then selects either a specified temperature or temperature gradient (heat flux). Then in the boxes to the right the numerical values of these quantities are entered in the form of a constant plus (possibly) a linear function of distance (X). In Figure 2, only the first three zones are in use; that is why only the first three rows of text input boxes have the white backgrounds.

Zone	Xmin	XMax	Temp	Grad				
1	0.0000	0.1575	○	○	T =	1.0	+ 0.0	X
2	0.1575	0.6575	○	○	dT/dy =	0.0	+ 100	X
3	0.6575	1.0000	●	○	T =	1.0	+ 0.0	X
4	1.0000	1.0000	○	○		0.0	+ 0.0	X
5	1.0000	1.0000	○	○		0.0	+ 0.0	X

Figure 2. Pop-up form for setting thermal boundary conditions.
Three of five possible zones have been used.

It should be remembered that the boundary layer equations themselves (Equations 1-3) and by implication these calculations, are not strictly applicable in the immediate vicinity of a sudden change in boundary conditions. Furthermore, the non-dimensionalization and interpretation of temperature results becomes somewhat hazy when a mix of temperature and temperature gradient conditions are specified along the plate.

Once all input parameters, including the **Reynolds Number** ($= U_\infty L / \nu$, where L is the plate length), fluid **Prandtl Number** ($= \frac{\nu}{\alpha} = \frac{c_p \mu}{k}$) and the freestream **Turbulence Intensity** (as a percent) have been specified, the user hits the **Plot** button and in a second or two the calculation, including the graphical output, is complete. Two hundred grid increments are used in the horizontal (flow) directions; at least 150 are used in normal direction, but as many as 450 more may be automatically added in the vertical where needed (e.g., for a low Prandtl number flow). The flow being parabolic, only a single sweep along the plate is used. Thus a typical run, even those which use the turbulence model, takes two seconds or less on a Pentium-level computer, including both the calculation and the plots. Warnings are issued for cases when calculations are clearly suspect, i.e., for cases involving very high or low Reynolds numbers and Peclet numbers (Pe = Re * Pr).

An Example Involving an Unheated Starting Length (Laminar)

In this section we discuss a typical application of this module and highlight the output. Input data corresponds to Problem 7.37 in the popular text by Incropera and Dewitt [10]. The plate Reynolds number is in the laminar range at 4.11×10^5, while the Prandtl number is taken as 0.71, corresponding to that of air at room temperature. The first two-thirds of the plate are unheated, while the rest is at a uniform temperature   Mixed thermal boundary conditions such as specified here are easily handled by the module (subject to the boundary layer approximations not being exactly satisfied in the immediate vicinity of sudden changes) and the computed results are displayed in Figure 1. A **Y-Scaling Factor** of 15 has been specified for this plot.

The gray area seen within the black window shows the extent of the computing region, which, using the transformed system of governing equations discussed earlier, grows along with the boundary layer. The white line indicates the outer edge of the velocity boundary layer, i.e., where the velocity reaches 99.5% of the freestream velocity. The color bands, which in this case do not start until two-thirds of the distance down the plate where the heating begins, show the computed isotherms. The thermal boundary layer is seen to be "catching up" with the growth of the velocity boundary layer - here both because the Prandtl number is somewhat less than 1.0 and because the thermal boundary layer is growing laterally into the lower velocity flow of a pre-existing hydrodynamic boundary layer instead of growing along with the velocity boundary layer into an undisturbed freestream.

The slider mechanism in the lower left hand corner may be used to "measure" both the surface temperature and the surface temperature gradient as a function of position along the plate. In this particular example, the surface temperature is either 0.0 or 1.0 and the surface temperature gradient is 0.0 for the first two thirds of the plate. All the data displayed using the scrollbar (surface temperature and temperature gradient as functions of x location) may be dumped to a text file for separate processing using the "Save Data" menu item. This text file may be readily imported into a spreadsheet for processing as discussed in the Appendix. For this unheated-starting-length problem, these "experimental" results would be compared with those using the traditional approach, i.e., using a correction factor derived using integral boundary layer methods [11] applied to the usual laminar flow results. The agreement is found to be excellent.

With the graphical display of Figure 1, it is easy to explain why the local heat transfer coefficient two-thirds of the way along the plate is higher with the unheated starting length than without. In the former case the thermal boundary layer is growing into an already-well developed velocity boundary layer. With the resulting lower velocities, it is able to diffuse normal to the plate more readily than otherwise.

An Example Involving Transition and Turbulence

In this section we consider a condition involving transition and turbulence. This run (Figure 3) involves a uniform plate temperature, Reynolds number of 1.5×10^6, Prandtl number of 1.0 and a freestream turbulence level of 1.0%, the latter corresponding to $Re_{x,tr} = 500,000$. As can be seen from the change in slope of the boundary layer edge, the transition to turbulence begins about a third of the way down the plate. To emphasize the transition process even more, we have crosshatched the plate itself from the computed point of transition to the position where the intermittency factor has reached 99%, the latter indicating a very nearly fully-turbulent flow. The standard correlations assume an abrupt jump from the local laminar value to the turbulent value that would exist had the boundary layer been turbulent all the way from the leading edge. Because of the use of the intermittency factors discussed earlier, our numerical predictions show a much less abrupt increase after transition and, in this case, a transition region extending nearly a third of the length of the plate.

4.1 Forced Convection on a Flat Plate

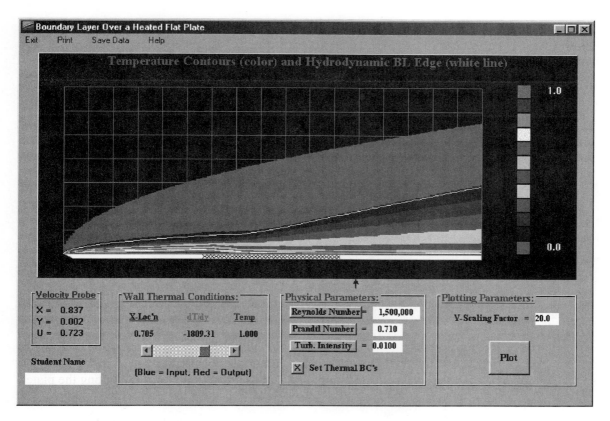

Figure 3. A run involving transition to turbulence. Transition begins 1/3 of the way down the plate at $Re_x = 500,000$. Since Pr ~ 1.0, the edges of the velocity and thermal boundary layers are nearly coincident.

If one traverses the length of the plate using the scrollbar, one finds a high value of the surface temperature gradient **dT/dy** (and by implication the convective heat transfer coefficient and local Nusselt number) at the leading edge, where the laminar boundary layer is thin. The value drops monotonically as the laminar boundary layer thickens up to the point of the onset of transition. Through the transition region, the value rises (though not instantly because of the effect of the intermittency factors used) to another high value. The interval during which the gradient is found to increase corresponds very closely to that indicated by the crosshatching. Then in the fully-turbulent region, the measured value again drops monotonically as the turbulent boundary layer grows thicker.

Verification of Results

The standard correlations for forced convection heat transfer from a flat plate may be used for verification of some of the results obtained from this simulation. The local Nusselt number for a wide range of Reynolds and Prandtl numbers is depicted in Figure 4. For laminar flows the Churchill and Ozoe correlation was used, while for turbulent flows the Chilton-Colburn correlation [10] was applied.

4.1 Forced Convection on a Flat Plate

Certain observations should be made. This module can be used, of course, for fluids, which do not exist, e.g., one having a Prandtl number of 0.3 as used in Figure 4. Since most correlations are based on experimental observations, none, including the Churchill and Ozoe correlation will claim a range of validity that includes that value since no fluids having a Prandtl number of = 0.3 exist.

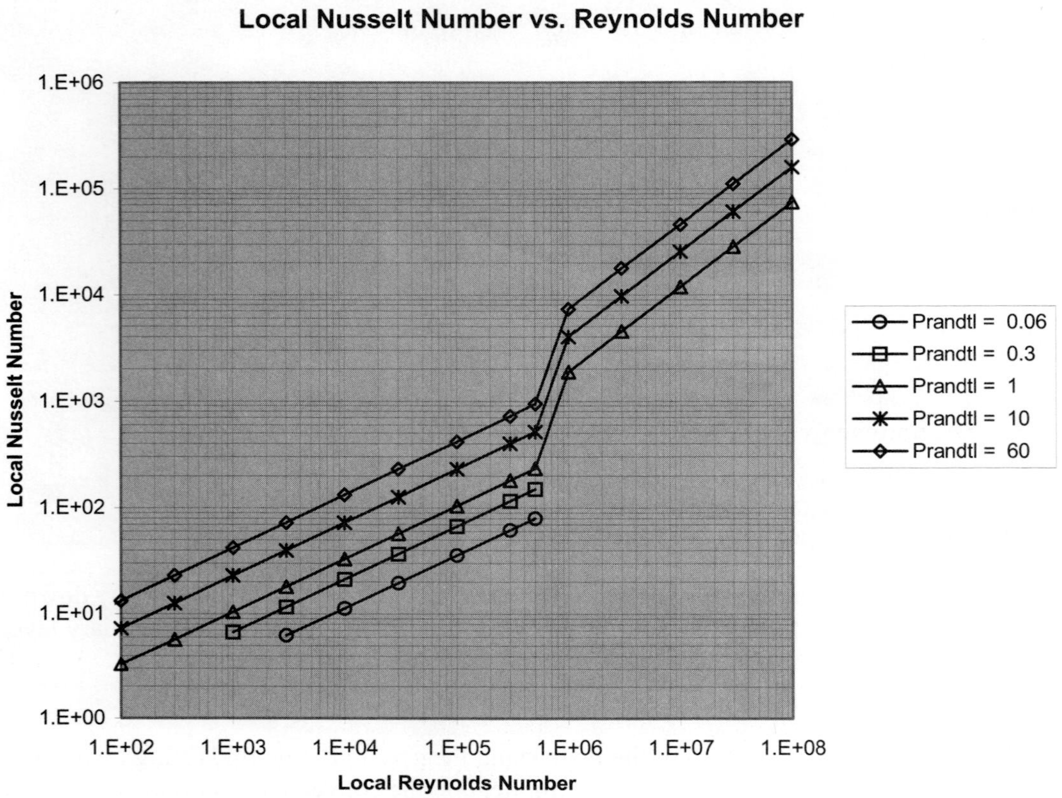

Local Nusselt Number vs. Reynolds Number

Figure 4. Local Nusselt number as a function of Reynolds and Prandtl numbers; results from standard correlations.

Convection heat transfer coefficients are usually presented in non-dimensional form and as correlations. As discussed also in the internal flows module, the Nusselt number itself is convenient, but misleading, since the values for low Prandtl number fluids (liquid metals) show lower Nusselt numbers than those for high Prandtl number fluids (oils), but the convection coefficients (h), the result of most interest to the engineer, are generally much higher in the former case.

NOTE

As of the date of the publication of this document, the module still uses the Anderson, et al. scheme for allowing the grid spacing to grow along with the boundary layer. While this method gives excellent agreement with correlations based on analytical solutions in the laminar region, the agreement is less satisfactory after transition to turbulence. Ideally the grid spacing should be refined again once the very steep gradients associated with transition and turbulent flows are reached. That weakness will be rectified in a future version of this code.

4.1 Forced Convection on a Flat Plate

References

1. Ribando, R.J., "Laminar Forced Convection on a Flat Plate", MAE 611 Course Notes, University of Virginia (1994).

2. Bejan, A., *Convection Heat Transfer*, Wiley, New York (1984).

3. Anderson, D.A., Tannehill, J.C., and Pletcher, R.H., *Computational Fluid Mechanics and Heat Transfer*, Hemisphere, New York (1984).

4. Ribando, R.J., Coyne, K.A., and O'Leary, G.W., "Teaching Module for Laminar and Turbulent Forced Convection on a Flat Plate", *Computer Applications in Engineering Education,* Vol. 6, No.2, pp. 115-125, 1998.

5. Anderson, E., Bai, Z., Bischof, C., Demmel, J., Dongarra, J., Du Croz, J., Greenbaum, A, Hammarling, S., McKenney, A., Ostrouchov, S., and Sorenson, D., *LAPACK Users' Guide*, Society for Industrial and Applied Mathematics (SIAM), Philadelphia (1992).

6. Schetz, J.A., *Boundary Layer Analysis,* Prentice Hall, Englewood Cliffs, NJ (1993).

7. Cebeci, T., and Bradshaw, P., *Physical and Computational Aspects of Convective Heat Transfer*, Springer-Verlag, New York (1988).

8. White, F.M., *Viscous Fluid Flow,* McGraw-Hill, New York (1974).

9. VanDriest, E.R., and Blumer, C.B., *AIAA Journal*, Vol. 1, pp. 1303-1306 (1963).

10. Incropera, F.P., and DeWitt, D.P., *Fundamentals of Heat and Mass Transfer*, 4th Ed. McGraw-Hill, New York (1996).

11. Kays, W.M., and Crawford, M.E., *Convective Heat and Mass Transfer*, 3rd Ed., McGraw-Hill, New York (1993).

Appendix: Calculation of the Nusselt Number

As discussed earlier, the interface is set up so that the user can "measure" both the surface temperature and the surface temperature gradient as a function of the x location. Both are reported, but of course in general one will be an input and the other an output depending on what the user has specified. (A color key is used to help the user keep track.) It is usually desirable to report either the local Nusselt number or a local heat transfer coefficient h_x. Thus in order to make the conversion we write:

$$q''(x) = -k_f \left.\frac{\partial T}{\partial y}\right|_{y=0} = h_x (T_w - T_\infty). \tag{A1}$$

We define non-dimensional temperatures and use the more conventional scaling for y, the normal coordinate, here:

$$T' = \frac{T - T_\infty}{\Delta T_{ref}}, \; y' = \frac{y}{L}, \; x' = \frac{x}{L}. \tag{A2}$$

4.1 Forced Convection on a Flat Plate

Using these in Equation A1, we get:

$$q''(x) = -k_f \frac{\Delta T_{ref}}{L} \frac{\partial T'}{\partial y'}\bigg|_{y=0} = h_x(T_w - T_\infty) \tag{A3}$$

Assuming we choose $\Delta T_{ref} = T_w - T_\infty$ then we have

$$-\frac{\partial T'}{\partial y'}\bigg|_{y=0} = \frac{h_x L}{k_f} = Nu_x \frac{L}{x}, \tag{A4}$$

or

$$Nu_x = -x' \frac{\partial T'}{\partial y'}\bigg|_{y=0} \tag{A5}$$

Here x' and $\frac{\partial T'}{\partial y'}\bigg|_{y=0}$ are the non-dimensional values read using the slider (or dumped to a file using the "Save Data" menu item). Now having computed Nu_x as a function of local Reynolds number (Re_x) and Prandtl number, one can find the local heat transfer coefficient from:

$$h_x \left[\frac{W}{m^2 K}\right] = Nu_x \, k_f \left[\frac{W}{mK}\right] \frac{1}{x[m]} \tag{A6}$$

Note that the local convective heat transfer coefficient h_x **is** directly proportional to the "measured" local surface temperature gradient. Thus, one can scroll along the surface and should observe that the convective heat transfer coefficient starts out high at the leading edge, decreases as the laminar boundary layer thickens, rises in the transition zone and then falls again as the turbulent boundary layer thickens.

Because all this can be confusing, let us give a numerical example. We can find the local Nusselt number corresponding to a (local) Reynolds number of 500,000 either by specifying 500,000 as the plate Reynolds number and taking data at the end of the plate or by specifying 1,000,000 as the plate Reynolds number and taking the measurement halfway down the plate (x/L = .5). (We use 500,000 as the upper limit for laminar flow at a freestream turbulence level of 1%.) The table shows how this works out.

Plate Reynolds #	Prandtl #	Turbulence (%)	$\frac{\partial T'}{\partial y'}\bigg|_{y=0}$ at Re_x = 500,000	x'	Nu_x
500,000	0.7	1.0	207	1.0	207
1,000,000	0.7	1.0	414	0.5	207

Both values compare very well with the value $Nu_x = 208$ given by the standard laminar flow correlation for this set of parameters. If one thinks in terms of the fluid properties and velocity

4.1 Forced Convection on a Flat Plate

being the same for both cases but the plate length being different, then the difference in numerical values returned for $\left.\frac{\partial T'}{\partial y'}\right|_{y=0}$ corresponds simply to the difference in scaling length used.

Heat Flux Boundary Conditions

Interpreting the numerical values that are returned and determining a Nusselt number for a case with mixed thermal boundary conditions may be difficult. Here we discuss a simple case involving a uniform heat flux, corresponding to a uniform $\left.\frac{\partial T'}{\partial y'}\right|_{y=0}$ applied along the entire length of the plate. Start again with Equation A1, this time noting that $T_w(x)$ is not constant:

$$q''(x) = -k_f \left.\frac{\partial T}{\partial y}\right|_{y=0} = h_x (T_w(x) - T_\infty). \quad (A7)$$

Scaling as above we get:

$$q''(x) = -k_f \frac{\Delta T_{ref}}{L} \frac{x}{x} \left.\frac{\partial T'}{\partial y'}\right|_{y=0} = h_x \Delta T_{ref} T'_w(x) \quad (A8)$$

This can be put into the form of a local Nusselt number:

$$Nu_x = \frac{h_x x}{k_f} = -\left.\frac{\partial T'}{\partial y'}\right|_{y=0} \frac{1}{T'} x' \quad (A9)$$

The three non-dimensional values on the right can all be read directly from the slider (or dumped to the file).

We can use the same parameters as before to give a numerical example for the heat flux boundary condition. This time we input a value for the non-dimensional surface temperature gradient. We choose somewhat arbitrarily a uniform value of −1000.0 (although this input value should be large enough in magnitude to give a significant temperature rise at the end of the plate) and run the same two cases as before.

Plate Reynolds #	Prandtl #	Turbulence (%)	$\left.\frac{\partial T'}{\partial y'}\right|_{y=0}$	x'	T' at $Re_x =$ 500,000	Nu_x
500,000	0.7	1.0	−1000	1.0	3.489	286.6
1,000,000	0.7	1.0	−1000	0.5	1.746	286.4

Both computed values for the local Nusselt number agree with the results from the accepted correlation for a uniform heat flux condition in laminar flow to within 1% [11]. Using these values (plus the fluid thermal conductivity) one can, for instance, easily determine the local surface temperature corresponding to a specified heat flux.

blproj.doc 12/20/00

4.2 Convective Heat and Mass Transfer from a Runner

Introduction

When the Olympics were held in Atlanta in the summer of 1996, the rejection of waste metabolic heat under extremely adverse conditions was a major concern to the athletes and their coaches. The human body, even that of a highly conditioned athlete, is a very inefficient "heat engine" and must reject large quantities of waste heat. In this exercise we will approximate the heat transfer from a scantily clad runner using a well-known correlation for an admittedly much simpler geometry, the cylinder in cross flow. Then, using the heat-and-mass transfer analogy, we will estimate the convective mass transfer and latent heat transfer due to sweating. With this model we can assess the role of the runner's velocity (relative to the wind), the skin temperature, ambient air temperature and relative humidity. As an additional exercise we will compute the "heat stress index" for several environmental conditions.

Figure 1. Sweaty Community Hero Olympic Torch Runner on Jefferson Park Avenue, Charlottesville, VA, June 21, 1996

Typically this problem would be done for a single set of parameters and left at that, but using a spreadsheet we are able to study a range of parameters readily and draw conclusions on the relative importance of each of them. Because these calculations require a number of fluid properties (five for air, six for water) over a range of temperatures, we have provided functions (macros) that can be invoked in exactly the same way as intrinsic functions (e.g., the sine, cosine and several hundred others in Excel), that is, directly from the cell formulae. We introduce the concept of "naming" cells, so that the resulting formulae are easily readable to the author and users of a spreadsheet.

Background

Since it would be unusual for the mechanical efficiency of the human body (defined as work accomplished (W) divided by metabolism (M) in the same units) to be more than 5 - 10% [1], getting rid of waste heat can be a major problem under hot and humid conditions. For most human activities the mechanical efficiency is nearly zero; under highly optimized conditions, e.g., a bicycle ergometer, the mechanical efficiency may reach just over 20% [2]. In this problem we consider in particular warm, humid situations, where getting rid of the balance (M - W) might pose a problem. At the other temperature extreme, the body will actually generate additional metabolic heat by shivering in order to maintain its temperature.

Waste heat is rejected both through respiration and by losses through the skin. We can write a *steady-state* energy balance for the body as [1]:

$$M - W = Q_{sk} + Q_{res}. \tag{1}$$

4.2 Convective Heat and Mass Transfer from a Runner

(This balance may be interpreted either on a total or per unit area basis.) The terms on the right may be further expanded to give:

$$M - W = (C + R + E_{sk}) + (C_{res} + E_{res}) \qquad (2)$$

Here the symbols C and R represent the convective (and conductive) and radiative sensible heat losses from the skin, respectively, while E_{sk} represents the evaporative (or latent) heat loss due to sweating. Sensible requires a temperature difference; evaporative or latent requires a concentration difference and phase change (evaporation). The terms C_{res} and E_{res} represent the convective (sensible) heat loss and the evaporative (latent) heat loss due to respiration. Air entering the lungs is heated from the ambient temperature to nearly the temperature of the body core, while it is expired with a relative humidity of nearly 100% based on the core temperature. Even though these terms may be sizable, we do not deal with respiratory losses here. We also neglect radiative losses (or gains on a sunny day), but will concentrate on the terms designated as C and E_{sk} in Equation 2.

A more refined two-node *transient* model is also discussed in the ASHRAE Fundamentals [1]. Here the body is divided into two concentric cylinders, the inner representing the core and the outer representing the skin layer. The core region nominally comprises about two-thirds of the body mass and its temperature is maintained at a constant temperature of 36.8 ± 1°C [3]. The core fraction does vary with ambient temperature as blood is kept within the core under cold conditions (vasoconstriction) and is pumped out into the skin layer under warm ones (vasodilatation). In the two-node model, heat stored in each is allowed to change with time, while heat transport between them is accomplished both by conduction through the tissue as well as by convection in blood. The results of even our simple steady-state model will show that such a transient model is needed in many cases. For instance, our results later show that under many conditions a sprinter cannot dissipate enough waste heat to maintain that pace indefinitely, and it is well known that athletes may run a fever of a few degrees under stressful conditions; in that case heat storage is evidently a factor. Though mammals, camels take advantage of thermal storage by actually allowing their core temperatures to drop by as much as 5 or 6°C by radiation to the clear nighttime sky, thus in effect storing up "coolness" to begin the next day. A typical camel conserves as much as 5 liters of water a day via "night setback" of his or her internal thermostat [4].

In this simple exercise we approximate the human runner as a cylinder in cross flow under steady-state conditions and use a standard forced convection correlation to estimate the heat and mass transfer under a variety of conditions. By automating the calculation a range of parameters may be easily studied and general trends noted.

Implementation

This exercise was originally designed to be implemented in a high-level programming language, e.g., Fortran 77/90 or C++, but in the implementation discussed here we will use a number of modern spreadsheet features to create a workbook that is easy to set up, understandable to the reader, easy to change and readily amenable to design studies. In particular, the use of Visual Basic for Applications (VBA) [5] from within Excel [6,7], means that those structured programming practices which normally are associated with high level programming languages can be used for the parts of the calculation where the complexity of the algorithm demands it, i.e., to avoid "spaghetti code," while the spreadsheet can be used for data input, simple calculations and tabular and graphical output. A brief introduction to VBA programming is given in the appendix on the CD-Rom.

4.2 Convective Heat and Mass Transfer from a Runner

Our estimate of the sensible and evaporative heat transfer from a runner involves a number of calculations [8]. The Excel workbook "Air-Water Properties" already has the needed fluid property functions (to be discussed later) included and is a good place to start. You need to set up a worksheet to study runner velocities ranging from 1.0 m/s to 10.0 m/s (the former corresponding to an "amble," the latter figure corresponding to a world class sprinter) in increments of 1.0 m/s. Major parameters to be studied, in addition to the runner's speed, are the ambient air temperature and relative humidity, as well as the skin temperature. Within fairly close limits this latter parameter is one that the body will regulate (along with the perspiration rate) in order to get rid of enough waste metabolic heat. Typically a value of 33.7°C (93°F) is taken as the skin temperature. A standard forced convection heat transfer correlation is to be used.

To begin, typical equivalent cylinder heights and diameters cited for an adult male are 1.8 m and 0.3 m, respectively, but the DuBois formula [9] may be used for others:

$$A_D = 0.108 m^{0.425} \ell^{0.725} \tag{3}$$

where,

A_D = DuBois surface area (ft^2)
m = mass (lb_m)
ℓ = height (in).

Almost each column on the spreadsheet after the first one containing the velocities requires one or more fluid properties, which for a single calculation would probably just be looked up manually in the tables provided in any heat transfer text. In this exercise, where we want to study a range of parameters conveniently, it is desirable to have all those properties easily accessible to the spreadsheet, either in tabular form or in the form of equations. So for instance, for each velocity from 1.0 to 10.0 m/s we first need to find the Reynolds number based on the runner's body diameter:

$$\text{Re}_D = \frac{\rho_{air} V \, Diameter}{\mu_{air}}. \tag{4}$$

The velocity (V) comes from the first column. The diameter comes from a particular cell, but rather than refer to that cell by its *absolute* address (using e.g., B9 to indicate an absolute instead of a *relative* address), we *name* the cell holding that value "Diameter" so that subsequent formulae clearly indicate that it is the diameter being used. Here the air properties density (ρ) and viscosity (μ) are to be evaluated at the *film* temperature, defined as the mean of the skin surface temperature and the ambient air temperature (both of which are input parameters). Fluid property functions written in VBA are provided and can be used directly in this formula. Thus, for example, the formula for cell B12 may read:

$$= Density\_Air(Tfilm) * A12 * Diameter / Viscosity\_Air(Tfilm). \tag{5}$$

Here *Tfilm* refers to the value stored in that named cell and "A12" is the cell address for the velocity for which we are computing the Reynolds number. By referring to three of the four variables in the above equation by meaningful names rather than cell addresses, this cell formula is vastly easier to understand and debug than otherwise. When copied to the remaining nine cells, only the row reference in the second factor should change.

4.2 Convective Heat and Mass Transfer from a Runner

The convective heat transfer coefficient ($\overline{h}$) for forced convection over a cylinder is given in non-dimensional form as a function of the Reynolds and Prandtl numbers:

$$\overline{Nu}_D = \frac{\overline{h}D}{k_{air}} = 0.3 + \frac{0.62\, Re_D^{1/2}\, Pr^{1/3}}{\left(1+(0.4/Pr)^{2/3}\right)^{1/4}} \left[1+\left[\frac{Re_D}{282{,}000}\right]^{5/8}\right]^{4/5}. \quad (6)$$

The former is computed in (5), the latter is another temperature-dependent fluid property. The calculation of this correlation of experimental data by Churchill and Bernstein [8] could probably be input into a particular range of cells (or because of its complexity would more likely be broken into a sequence of operations and spread across several columns), but it is probably easier to open up a module *(Tools, Macro, Visual Basic Editor)* and then write a nicely structured and documented VBA function to evaluate the expression. The Reynolds number from the second column and the Prandtl number from above are passed into the function, and the value of Nusselt number based on diameter is returned - just as they would be in any high-level programming language. Thus, a typical cell formula invoking this user-supplied Nusselt function might read: *=Nusselt(B12,Prandtl)*. From the Nusselt number, the convective heat transfer coefficient can be determined as:

$$\overline{h} = \frac{\overline{Nu}_D}{Diameter} k_{air} \quad (7)$$

Here again, the thermal conductivity of air (k_{air}) is one of the property values provided as a VBA function. The sensible heat transfer follows directly from the convective heat transfer coefficient, the cylinder surface area (neglect the ends) and the driving temperature difference T_{skin} - T_{air}. At this point you probably have used five columns: velocity, Reynolds number, Nusselt number, convective heat transfer coefficient and sensible heat transfer.

Now using the analogy between heat transfer and mass transfer, estimating the latent or evaporative heat loss is a relatively easy matter. At first we assume that the body (cylinder) is entirely covered with sweat. Now feed the Reynolds number and the Schmidt number ($= \frac{\mu_{air}}{\rho_{air} D_{ab}}$) to the same Churchill and Bernstein function, and this time it will return the Sherwood number ($= \frac{h_m\, Diameter}{D_{ab}}$). Here D_{ab} represents the binary diffusion coefficient for water vapor in air (and is a tabulated number). With the Sherwood number you can solve for the convective mass transfer coefficient (h_m).

The convective mass transfer is determined from the product of h_m, the runner's surface area and the concentration difference between the surface and the ambient air. The latter is based on an assumed 100% relative humidity at the skin surface temperature and the specified ambient value away from the surface based on temperature and relative humidity. Of course, if the runner sweats too profusely, much of it will just drip off without evaporating. The ASHRAE Handbook [1] gives an approximate upper limit of 4.0 lb$_m$/hr (1.8 kg/hr) for the total sweat rate for an average male, and a maximum useful rate of approximately 2.4 lb$_m$/hr (1.1 kg/hr). Finally, knowing the heat of vaporization, the latent heat transfer can be found on the assumption that the body is providing the heat necessary to evaporate all the sweat. The properties of water needed, including the specific volume of water vapor and its heat of vaporization (h_{fg}), are again provided as VBA functions. (In fact, the only property not provided by the functions is the binary diffusion coefficient.)

4.2 Convective Heat and Mass Transfer from a Runner

Some sketchy data for the metabolism (M) as a function of velocity based on the hypothetical dimensions given above are given in Lunardini [10]:

Velocity (m/s)	Metabolism (Watts)
.67	188
1.34	273
1.8	378
4.5	1050
10.0	4200 (estimated)

The rate while sitting quietly is about 100 W. Some similar data for bicyclists may be found in [2]. At the lower velocities we would conclude that the runner would be able to dissipate all the heat necessary; his skin wettedness would simply be less than the 100% assumed in our calculation. Your calculations will show that even at a relatively low humidity level and comfortable air temperature a sprinter would have difficulty getting rid of enough heat, and thus we would conclude that such a pace could not be kept up very long - merely from a heat transfer standpoint.

Task List

1. Create spreadsheet as discussed.

2. Run your spreadsheet for the parameters given below (and for your geometry) and then plot the sensible, latent and total heat transfer for velocities from 1.0 to 10.0 m/s in 1.0 m/s increments.

    Case A: $T_{sk} = 34$ °C (93 °F), $T_{air} = 24$ °C (75 °F), R.H. = 50%
    Case B: $T_{sk} = 34$ °C (93 °F), $T_{air} = 32$ °C (90 °F), R.H. = 80%

    In this part you are assuming that the whole body cylinder is covered in sweat, so your predictions are of the maximum possible heat the runner could get rid of under these conditions and not what he or she would actually generate. Compute and report the sweating rate for all cases and compare with the values from the ASHRAE handbook mentioned earlier.

3. Start a new worksheet and for the fourth case in Lunardini's data in the notes you are to create a contour map of the heat stress index as a function of temperature (from 280 - 315 K in increments of 5 degrees) and relative humidity (from 0.0 to 1.0 in 0.1 increments). Now you will have to create one grand VBA function that accomplishes everything you did with cell formulae in Step 2. The percent wettedness comes from: total heat to be transferred (1000 W here) = sensible heat transfer + % wettedness * latent heat transfer based on 100% wettedness (the quantity computed in Step 2). This number should be limited to 100%.

Conclusion

Obviously the model implemented here is extremely simple. It can, however, at least be used to explain a number of commonly observed phenomena, e.g., the very big effect of ambient humidity on one's comfort level, that an athlete's stamina is higher on the open road than it is on a stationary bicycle or treadmill in a stuffy gym, etc. Of course, anyone can come with some objections to the assumptions made. A more sophisticated model that addresses a number of those issues has been presented by Galbis-Reig [11]. Additional current references may also be found there and in Reference [12]. Recent

4.2 Convective Heat and Mass Transfer from a Runner

research involving enhancing athletic performance through artificial removal of waste metabolic heat through the hands is discussed in [13].

References

1. *1993 ASHRAE Handbook - Fundamentals Handbook*, American Society of Heating, Refrigerating and Air-Conditioning Engineers, Inc., Atlanta (1993), Ch. 8.

2. Whitt, F.R. and Wilson, D.G., *Bicycling Science*, MIT Press, Cambridge, MA (1982).

3. Cooper, K.E., "Body Temperature Regulation" in *Encyclopedia of Human Biology*, R. Dulbecco, Vol. 1, Academic Press, (1991).

4. Curtis, H., *Biology*, 2nd Ed., Worth Publishers, New York (1975).

5. Walkenbach, J., *Excel for Windows 95 Power Programming with VBA*, 2nd Ed., IDG Books Worldwide, Inc., Foster City, CA (1996).

6. Etter, D.M., *Microsoft Excel for Engineers*, Addison-Wesley Publishing Company, Menlo Park, CA (1995).

7. Gottfried, B.S., *Spreadsheet Tools for Engineers - Excel 97 Version*, McGraw-Hill, New York (1997).

8. Incropera, F.P., and DeWitt, D.P., *Fundamentals of Heat and Mass Transfer*, 4th Ed., Wiley, New York (1996). (This project originated with problem 6.53.)

9. Dubois, D., and Dubois, E.F., "A Formula to Estimate Approximate Surface Area if Height and Weight are Known", *Archives of Internal Medicine*, Vol. 17, pp. 863-71 (1916).

10. Lunardini, V.J., *Heat Transfer in Cold Climates,* Van Nostrand Reinhold, New York (1981).

11. Galbis-Reig, V., *Computational Modeling of the Human Thermal Regulatory System*, B.S. Thesis, University of Virginia (1997).

12. Çengel, Y. A., *Heat Transfer, A Practical Approach,* WCB McGraw-Hill, Boston (1998), pp. 706-711.

13. Guterman, L., "Putting the Power of Healing in the Palm of Your Hand," The *Chronicle of Higher Education*, June 9, 2000.

4.2 Convective Heat and Mass Transfer from a Runner

Appendix: Implementation of the Property Functions

Visual Basic for Applications (VBA) provides a wealth of programming structures and features [5]. Here we focus specifically on the property functions provided for use with this project. State and transport properties of air needed in this calculation are the thermal conductivity, constant pressure specific heat, Prandtl number, density and viscosity, all at atmospheric pressure. The same transport properties plus the specific volume of the saturated vapor and latent heat of evaporation of water are also needed for the mass transfer (sweating) calculation. While Excel does provide functions for table lookups [6], they do not linearly interpolate for values not explicitly listed in the tables. Thus a function was developed for each of the needed air and water properties. The user simply invokes these functions in the same way he or she would use a supplied, e.g., trigonometric, function. The eleven user-invoked functions, each of which goes by a highly descriptive name (Prandtl_Air, Viscosity_Water, etc.), in turn invoke a single interpolation routine. Arguments sent from the property functions to the interpolation function are the interpolation variable (temperature in all cases), the column reference in the data table for the particular property and the identifier of the fluid. A message box warning that the sought-after temperature value is outside the range of the table is returned where appropriate.

Each of the functions listed below returns the requested air or water property as a function of temperature in degrees Kelvin. The units for each are given at the top of the VBA function itself.

Conductivity_Air(Temp)	Conductivity_Water(Temp)
Cp_Air(Temp)	Cp_Water(Temp)
Prandtl_Air(Temp)	Hfg_Water(Temp)
Density_Air(Temp)	Prandtl_Water(Temp)
Viscosity_Air(Temp)	Viscosity_Water(Temp)
	SpVolVap_Water(Temp)

Runner.doc 6/8/00

5.1 Forced Convection - Internal Flows

Introduction

This module might be considered a *virtual* laboratory; it is based somewhat on an experiment the author performed many years ago in undergraduate thermal sciences lab. The initial length of the pipe (Figure 1) is not heated; in this *hydrodynamic entry region* the velocity profile becomes fully developed. Immediately downstream is a section of the pipe where a sudden change in the wall thermal boundary condition is introduced. That condition might involve a change in wall temperature to a uniform value different than that of the entering fluid or it might involve a prescribed, uniform heat flux into or out of the fluid. The subsequent development of the temperature field in the fluid is the focus of this module and is, of course, the underlying physics behind the standard analytically and experimentally determined internal flow correlations.

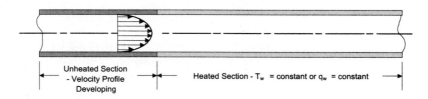

Figure 1. Schematic of circular pipe

The Fortran "engine" running behind this module began as a project in a graduate-level computational methods course. Besides having the analytical solution for laminar flow due to Graetz, Nusselt and others available for comparison [1], this problem has some interesting numerical challenges due to the cylindrical geometry [2]. Following the step change in wall thermal condition, a single partial differential equation describes the subsequent development of the temperature distribution and far enough downstream, the temperature distribution and the bottom line, the Nusselt number, asymptote to the fully developed condition. More recently, using what is know as a "mixing length" model, we extended the numerical algorithm to include transition and turbulent flows. You may recall that for turbulent flows the *thermal entry length* [3] is supposed to be nearly independent of Prandtl number and only on the order of 10 - 60 diameters in length (vs. $Re \cdot Pr / 20$ for laminar flows), but solving the equations as a thermal entry length problem turns out to be very convenient computationally. The table below shows the applicability of the calculations performed by this module. The solution of the combined entry length problem, while not impossible, would require the solution of the momentum and continuity equations, which are not currently included.

	Laminar	Transition	Turbulent
Thermal Entry Length	yes	yes	yes
Fully-Developed	yes	yes	yes
Combined Entry Length	no	no	no

For a turbulent flow the single governing energy balance equation is written [4]:

5.1 Forced Convection – Internal Flows

$$\rho c_p u \frac{\partial t}{\partial x} = \frac{k}{r} \frac{\partial}{\partial y}\left[r\left(1 + \frac{Pr}{Pr_t}\varepsilon_m^+\right)\frac{\partial t}{\partial y}\right] \quad (1)$$

Here y is conventionally measured from the wall, while r is measured from the centerline of the pipe. Lower case t represents a dimensional value. This equation expresses the balance between axial transport of heat by advection, i.e., bulk fluid motion (on the left where ρc_p is the heat capacity per unit volume) and radial transport by molecular conduction and turbulent mixing processes (on the right, where k is the thermal conductivity). Other than the addition of the extra term representing turbulent transport and the fact that the velocity profile is not a simple parabola as it is for laminar flows, but is determined from an empirical *mixing length* model, this equation and the procedures used to approximate it are similar to that used earlier for the laminar, thermal entry length problem. The exact details of how we model the turbulent Prandtl number (Pr_t) and the eddy diffusivity (ε_m^+) are beyond the scope of this write-up [5-7]. Suffice it to say that the extra term represents an additional transport mechanism besides molecular conduction for cross-stream transport, thereby enhancing the convective heat transfer. The program automatically includes an approximate model for these terms once the Reynolds number based on diameter exceeds 2300, the value usually used as the criterion for transition to turbulence in a cylindrical pipe. In contrast to the external flow module where the transition to turbulence is dramatic, here there is no definitive indication directly visible on the screen. Furthermore, because the friction factor is used in computing the mixing length model parameters, the fact that the heat transfer results from this model do match fairly well with the usual convection correlations [3] is testimony to the validity of the well-known analogy between heat and mass transfer - rather than representing a computational triumph.

The numerical computations taking place "behind the scenes" in this module are considerably simpler than what was solved in the external flow module. There the boundary layer forms of the continuity, x-momentum and energy equations were being approximated. For this thermal-entry length, internal flow problem, the velocity profile is assumed not to change and only the energy equation (Equation 1) is being approximated. That equation is mathematically parabolic - just like a transient conduction equation. Here, however, the equation is parabolic in space rather than in time, so this solution is "marched" forward in space (down the pipe), rather than in time [2]. At each axial position along the pipe, a control volume energy balance, which when that volume is reduced to infinitesimal size corresponds to Equation 1 above, is written. (Note that other than at the centerline, the small control volumes used here are toroidal.) Derivatives in the radial direction are expressed in *implicit* form, so that the energy balance yields a tridiagonal system of linear equations that is solved at that particular station before moving on to the next. When this program was originally developed on a much slower computer, the marching process was very evident, but now it is much too fast to see.

Bear in mind that all fluid properties, including density, specific heat, thermal conductivity and Prandtl number, are assumed to be uniform and unchanging in Equation 1. The turbulent Prandtl number (Pr_t) and eddy diffusivity (ε_m^+) are functions of the radial position only. Including the effects of density and viscosity changes would necessitate the solution of the continuity and axial momentum equation in tandem with the energy equation, which is clearly possible, but has not been implemented.

5.1 Forced Convection – Internal Flows

Program Operation

User inputs are required in all the white boxes in the bottom right area of the interface (See Figure 2). These include the **Reynolds Number** based on diameter, **Prandtl Number** of the fluid and the **Length-to-Diameter** ratio of the cylindrical pipe. For almost any practical L/D a scaled drawing of the pipe would be very long and narrow, so the **Plot Scale** entry allows the user to expand the radial dimension relative to the length (up to a certain limit). In Figure 1, for instance, the radial dimension has been expanded by a factor of 13.5, thus allowing the user to better see the temperature and heatline distribution.

Just to the left of center in the bottom half, option (radio) buttons allow the user to specify the **Wall Conditions** (type of heating) and the **Heating Direction**. **Constant** (wall) **Temperature** implies just that; perhaps steam is condensing at constant pressure on the outside surface of a pipe made of a thin, highly conducting material - thus keeping the inside surface temperature constant. **Constant Gradient** indicates that the heat flux into the fluid is uniform. In practice this condition might be achieved by wrapping the pipe uniformly with an electric heating element. This heating element would be wrapped in turn with insulation. The assumption is that there is no axial conduction; any heat generated at a particular position along the pipe has nowhere to go but through the pipe wall and into the passing fluid at that location. Although not included in this model, the use of electric heating wires does invite the testing of heat flux distributions other than strictly uniform. The spacing of the electric heating wires or the thickness of the elements themselves can be tailored to give the desired distribution. The **Heat Direction** button has no effect at all on the calculation, but does affect the color contour plot and the heatlines. With the **Hot Wall** option the entering fluid is colored blue while the most heated fluid shows as red. The reverse is true for **Cold Wall**.

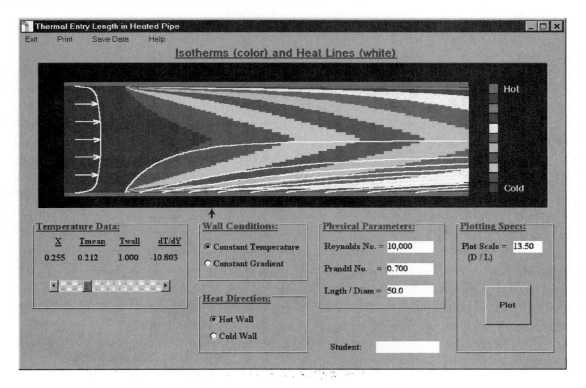

Figure 2. Interface showing results for a constant wall temperature case

5.1 Forced Convection – Internal Flows

Clicking the **Plot** button in the lower right corner causes the calculation to run using the parameters as set.

Program Output and Interpretation

Program output comes in three forms: the graphical display in the upper part of the interface, the slider bar display in the bottom left corner and an optional data file that may be generated. The slider bar returns as a function of position along the pipe the mean temperature (**Tmean**), the wall temperature (**Twall**) and the temperature gradient at the wall (**dT/dY**). Since the length of the pipe has been cut into 200 increments for this calculation, the slider steps in increments of .005. The mean temperature increases or decreases according to the direction of heat flow and as noted later is a direct measure of the amount of heat flowing by a particular axial station. For a constant gradient condition, the mean temperature should increase or decrease linearly with X. For a constant wall temperature case, the value of Tmean at the exit can be used to find the *effectiveness* (= q/q_{max}) of the whole heat transfer process.

For a constant wall temperature case, the value of Twall will stay at either 0.0 or 1.0 depending on the heat flow direction along the length of the pipe and will adjust as needed to maintain the desired heat flux in a constant gradient case. Similarly, the temperature gradient at the surface (dT/dY) will be uniform for the constant gradient case (where it is, in effect, an input) and a function of X for the constant temperature condition (where it is an output). All this data may be saved in a single data file using the **Save Data** menu option and imported into, e.g., a spreadsheet. Instructions for processing this data to find the convective heat transfer coefficient and Nusselt number are contained in the following section. Values of T_m and T_w when plotted as functions of distance along the pipe display the behavior expected for constant wall temperature and constant heat flux conditions.

The graphical display at the top shows the **Velocity Profile**, **Isotherms** and **Heatlines**. Flow is assumed from left to right. A plot of the radial distribution of velocity is shown at the left side just upstream of the start of the heated section. Since this is a thermal entry length problem, that velocity profile is assumed to be fully developed before entering the heated zone and does not change during its traverse of the heated length, i.e., effects of property changes, e.g., viscosity, with temperature are not considered. For laminar (Poiseuille flow) the velocity profile is seen to be a parabola; at higher Reynolds numbers the profile is seen to flatten out as expected. The computed temperature field is shown by the colored isotherms; being axisymmetric the isotherms showing above the centerline are mirror images of those below. Many workers in this field present only isotherms; we have superimposed *heatlines* over the lower half of the isotherms (providing the display has been expanded enough to allow them to be shown clearly). For situations involving only conduction heat transfer, isotherms make a useful display since the heat flow is always normal to them, but that is not the case when there is fluid motion. Bejan [4] suggests heatlines as an alternative presentation means. Since they are the analog of streamlines, heat flows parallel to the plotted heatlines by both bulk fluid motion (in this problem in the axial direction only) and diffusion (here including conduction and turbulent transport in the radial direction only). By convention heatlines are calculated relative to the lowest temperature in the flow.

The heatlines and isotherms may be used together to understand what is going on physically within this flow. In Figure 2 one notes that the sidewall boundary layers merge at the centerline only about 20% of the way down the pipe. We note that a single isotherm runs along the inside surface of the pipe (consistent with this being a constant wall temperature case), and that the temperature gradient along the wall does not appear severe. In addition the heatlines are

5.1 Forced Convection – Internal Flows

seen to emanate from the wall (consistent with our heating a cold fluid by contact with a hot wall). Observe that immediately adjacent to the wall the heatlines are directed radially - cross-stream transport by molecular conduction dominates near the wall where the axial velocity is low. Further out where most heat transport is by bulk flow in the axial direction, they are nearly parallel to the flow. The spacing of the heatlines as they leave the wall is not uniform. More heat is transferred to the fluid immediately at the beginning of the heated length where both the heat transfer coefficient and the local temperature difference between wall and fluid are greatest. The cylindrical geometry does complicate the interpretation of the heat flow lines. Observe that at the outlet, the spacing between lines is bigger near the centerline. In the integral used to find the mean temperature (See next section), note that while the velocity is the highest near the centerline, the flow area associated with a given increment in radius is smaller than where closer to the wall.

Figure 3 shows another run with a quite different set of parameters. The Reynolds number is higher than in the first sample, so the velocity profile is flatter. Since this case involves a prescribed constant temperature gradient, the slider bar shows a value of dT/dY = 1.0 at the particular location where it happens to be set (as well as along the whole length). The radial temperature gradient along the wall is noticeably steeper than in Figure 2. Moreover the temperatures adjacent to the wall increase along the flow direction to maintain the prescribed heat flow. (The increasing wall temperature can be seen using the slider bar window as well.)

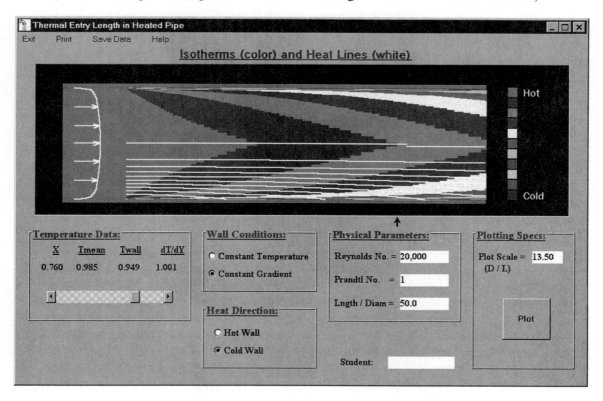

Figure. 2. Interface showing results for a constant heat flux case

5.1 Forced Convection – Internal Flows

Since in this run the temperature of the entering fluid is higher than that of the wall, ten full increments of heatlines are evident at the start of the heated length. Again the spacing of the heatlines seen entering the wall is uniform in the axial direction, but only 35-40% of the incoming heat (measured relative to the lowest wall temperature (a variable in this problem) is transferred to the wall.

Note that while this model does not include any attempt to model the effects of heat transfer enhancement devices [8], careful study of the graphical output as seen in Figures 2 and 3 can certainly uncover those cases where use of such devices might prove beneficial.

Using the Results/Finding the Nusselt Number

As noted earlier the "Save Data" menu will allow the user to save the wall temperature, fluid mixed mean temperature and the value of the temperature gradient at the surface in a text file, which may be imported into a spreadsheet for further processing.

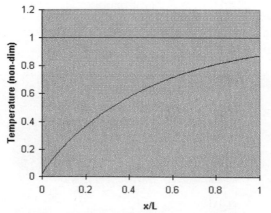

Figure 4. Wall temperature and fluid mean temperature for Re_D = 15,000, Pr = 0.7, L/D = 150, constant wall temperature

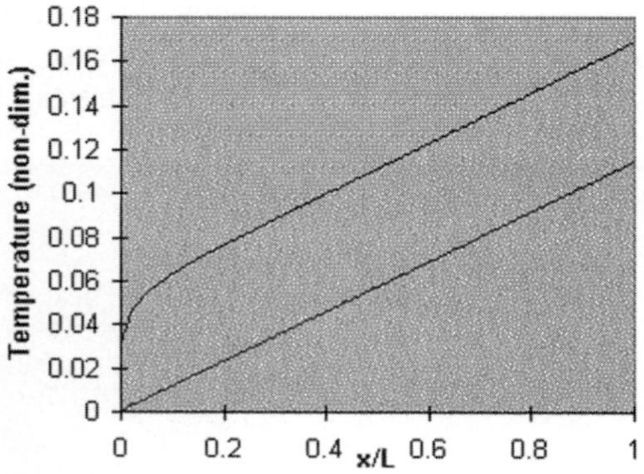

Figure 5. Wall temperature and fluid mean temperature for Re_D = 15,000, Pr = 0.7, L/D =150, constant surface heat flux

Figures 4 and 5 show typical data that was saved and plotted for identical parameters except that a constant wall temperature thermal boundary condition was applied in Figure 4, while a constant heat flux was applied in Figure 5. (The calculation of the fluid mixed mean temperature is discussed later.) Note that as expected the temperature difference in Figure 4 decays exponentially along the length of the pipe. In Figure 5 the fluid mean temperature increases linearly along the pipe, as it must; while, after a very short thermal development length, the wall temperature also increases linearly, indicating then a fully developed and constant heat transfer coefficient. The Figure 4 case probably has a short thermal development length also, but that is not so readily seen.

Since the calculation is done over a finite-length of pipe, one can use this module quite successfully for a myriad of internal flow problems without even worrying about heat transfer coefficients or correlations. For instance for the Figure 4 cases above one would merely have to multiply the mean exit temperature seen in the graph (~87%) by the actual approach temperature difference $(T_s - T_i)$ to get the exit temperature. The effectiveness for this case would be about 87%.

5.1 Forced Convection – Internal Flows

For the constant heat flux case one can use the actual heat added to or removed from the fluid over the specified length to find the actual change in mean temperature. The computed non-dimensional mean outlet value (.114 in Figure 5) then corresponds to that change. Knowing that scaling then one can determine the surface temperature anywhere along the pipe.

If one needs them (and one certainly <u>should</u> do this to verify these numerical results), there are several ways to use the computed results to determine the Nusselt number (and thus h, the convective heat transfer coefficient). One of them involves use of the surface temperature gradient; this method as well as the second requires knowledge of the mixed mean temperature:

$$t_m(x) = \frac{2}{u_m r_o^2} \int_0^{r_o} u\, t\, r\, dr \qquad (2)$$

Remember that $t_m(x)$ is a direct measure of the amount of thermal energy flowing past a particular x station since it is weighted by both the velocity and flow area. Early in this development we had envisioned the user "measuring" the radial distribution of velocity and temperature at several locations using simulated instrumentation, then doing a numerical integration, but realized that that would have taken too much class time. Thus the quantity $t_m(x)$ (or more precisely the non-dimensional value $T_m(x)$) is found within the program using a numerical integral of the computed results at each axial station along the pipe. The results may be recorded manually using the slider mechanism in the lower left corner, or the entire distribution may be saved in a text file that then can be imported into a spreadsheet for post-processing.

Then we write:

$$q = h(t_s - t_m) = + k_f \left.\frac{\partial t}{\partial r}\right|_{r=r_o} \qquad (3)$$

that is, the heat transfer can be found by evaluating the temperature gradient in the fluid at the surface. Now nondimensionalize the radial coordinate by the radius of the tube (r_o) and temperatures by the difference between the wall temperature and the entering fluid temperature:

$$T = \frac{t - t_i}{t_s - t_i} \qquad (4)$$

For heating cases involving a fixed wall temperature, this scaling gives $T_s = 1.0$ and the inlet temperature is $T_i = 0.0$. These numbers will be reversed for fluids being cooled. Then substituting in (3), we get:

$$h(T_s - T_m) = + \frac{k_f}{r_o} \left.\frac{\partial T}{\partial r}\right|_s \qquad (5)$$

Rearranging yields:

$$\frac{hD}{k_f} \equiv Nu_D = \frac{+2 \left.\frac{\partial T}{\partial r}\right|_s}{T_s - T_m} = \frac{-2 \left.\frac{\partial T}{\partial y}\right|_{y=0}}{T_s - T_m} \qquad (6)$$

5.1 Forced Convection – Internal Flows

The slider mechanism will give the non-dimensional values of the wall temperature (T_s), mixed mean temperature (T_m) and the temperature gradient (as $\left.\frac{\partial T}{\partial y}\right|_{y=0}$) at the wall directly.

The second approach to finding the convective heat transfer coefficient is to imagine a control volume consisting of a short length of the pipe and say that the heat added through the sides of that volume should be equal to the difference between what is transported through the upstream face and that which flows out the downstream face:

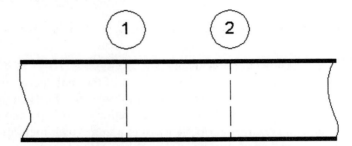

Figure 6. Control Volume for Nusselt Number Calculation

Remembering that t_m as defined in (2) above is the appropriate mean to use for the latter (advected) quantities, we can write that heat balance for this short length symbolically as:

$$q = h(x) 2\pi r_o (t_s - t_m) \Delta L = \dot{m} c_p (t_{m2} - t_{m1}) = \rho \overline{V} \pi r_o^2 c_p (t_{m2} - t_{m1}) \qquad (7)$$

This balance can be rearranged to give:

$$\frac{h}{\rho \overline{V} c_p} = \frac{Nu}{Re\, Pr} \equiv St = \frac{r_o (T_{m2} - T_{m1})}{2\Delta L (T_s - T_m)} \qquad (8)$$

where now we have switched to the non-dimensional temperatures and St represents the Stanton number. The T_m in the driving temperature difference in the denominator can probably be approximated as the mean of the upstream (T_{m1}) and downstream (T_{m2}) values. The same applies to the wall temperature T_s in the case of a heat-flux-specified condition (where it will be a function of x). The length of the section of pipe ΔL should be long enough so that the change in mean temperature is measurable, but not too long that the change can not be considered linear. It can be found in terms of the overall length of the pipe using the slider mechanism.

A Very Important Note on Accuracy!

Just as poor experimental practices in a laboratory will generally yield poor results, poor technique on the part of the user can give poor results in this simulated experiment. As an example consider a constant wall temperature case. If you use an L/D ratio so high that the heat transfer is nearly complete in the first quarter of the pipe length, then a Nusselt number computed further along the pipe using a Tmean and Twall very close to each other will be highly suspect.

Also, while the **Save Data** option conveniently allows you to import all the scrollbar data into a spreadsheet and find a Nusselt number anywhere along the length (this capability is very convenient for analyzing the thermal entry length problem in particular), you should bear in mind

5.1 Forced Convection – Internal Flows

that there are but 200 axial grid spacings. A local Nu_D based on the computed values at the fourth axial computational plane will only have three grid points used upstream of it (and this being a one way calculation, 197 downstream points being wasted) and will be correspondingly inaccurate. Thus to obtain the best local values, the recommended protocol is to make the pipe L/D value appropriate to that position (thus having 200 grid spacings upstream of the point of interest) and use the data generated right at the end of the pipe.

Verification of Results

Nearly all of the results generated by this program may be verified by direct comparison to the standard correlations. Note that the program can easily simulate non-existent fluids! An Excel^{tm} spreadsheet that implements many of the usual correlations may be found on the CD and is discussed in the next writeup.

References

1. Kays, W.M. and Crawford, M.E., *Convective Heat and Mass Transfer*, 3rd Ed., McGraw-Hill, New York (1993).

2. Ribando, R.J., and O'Leary, G.W., "Numerical Methods in Engineering Education: An Example Student Project in Convection Heat Transfer," *Computer Applications in Engineering Education,* Vol. 2, No. 3, 1994, pp. 165-174.

3. Incropera, F.P., and DeWitt, D.P., *Fundamentals of Heat and Mass Transfer*, 4th Ed., Wiley, New York (1996).

4. Bejan, A., *Convection Heat Transfer*, 2nd Ed., Wiley-Interscience, New York (1995).

5. Schetz, J.A., *Boundary Layer Analysis,* Prentice-Hall, Englewood Cliffs, NJ (1993).

6. Cebeci, T. and Bradshaw, P., *Physical and Computational Aspects of Convective Heat Transfer,* Springer-Verlag, New York (1988).

7. Ribando, R.J., "Turbulent Pipe Flow", MAE 611 unpublished course notes, University of Virginia (1993).

8. Webb, R.L., *Principles of Enhanced Heat Transfer*, Wiley, New York (1994).

5.2 Internal Flow Correlations

While developing the internal flow module discussed in the previous section, we found it convenient to create a workbook capable of returning benchmark values from conventional internal flow convection correlations. That Excel[tm] workbook, which implements the particular set of correlations recommended in one popular textbook [1], may be found on the CD-ROM. Perhaps just as useful as the numerical values of the Nusselt number based on diameter (Nu_D) which that workbook generates is the Visual Basic for Applications (VBA) module accompanying the spreadsheet. That program code clearly shows the logic used in selecting an appropriate correlation for the indicated situation. A flowchart for the VBA module is presented in Figure 1 below. Each of the correlations has some recommended range of applicability and, while not shown in complete detail on the flow chart below, all the restrictions are enforced in the module through conditional (IF) statements. Note that there are two correlations due to Sieder and Tate: one for the laminar combined entry length problem (velocity and temperature profiles developing simultaneously), while the other is for fully turbulent flows characterized by large property variations.

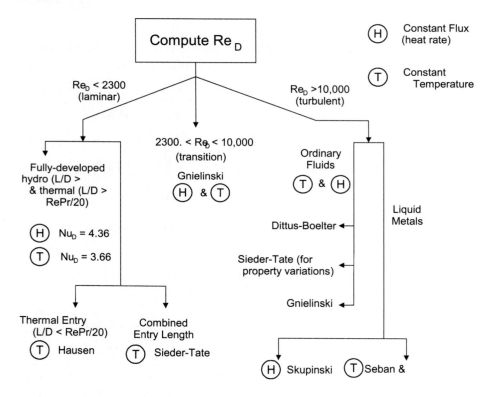

Figure 1. Flow chart for selecting internal flow correlations

Inputs to the function are the Reynolds number based on diameter, Prandtl number of the fluid, Length/Diameter of the pipe, a flag to indicate whether a constant temperature or constant heat flux wall condition applies, the friction factor and, finally, a flag which asks that the function return a textbox that explains what correlation was used in computing the Nusselt number it returns or why none was found to be applicable. Whether the thermal condition is one of constant temperature or constant heat flux does not matter for the turbulent flows other than those of liquid metals. The friction factor is only needed for turbulent flows in non-smooth tubes. Results spanning a wide range of Prandtl and Reynolds numbers are plotted in Figure 2 and are

5.2 Internal Flow Correlations

representative. A length/diameter ratio of 100 was used for all cases shown on this plot. That value is sufficiently large that all cases, both laminar and turbulent, may be taken to be fully developed hydrodynamically (velocity profile fully developed). (For laminar flows $(x/D)_{\text{fully-developed}} > Re/20$; for turbulent flows the L/D range is commonly taken as from about 10 to 60.)

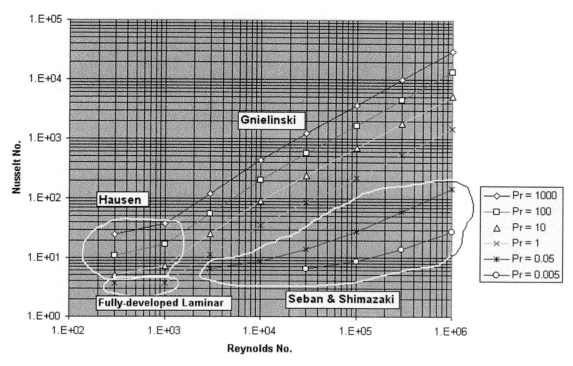

Figure 2. Typical spreadsheet results for Nusselt number (L/D = 100), Constant Wall Temperature

For the liquid metals (Pr = .05 and .005) the Seban and Shimazaki correlation was found applicable for the higher Reynolds number (turbulent) cases, but because of its lower limitation on Peclet number (Pe = Re*Pr > 100) cannot be used for the lower Reynolds number (all laminar and several turbulent) liquid metal cases. For Pr = 1 and both laminar Reynolds number cases the condition that the flow be fully-developed thermally (L/D > RePr/20) is met, and thus the constant value of 3.66 is returned for those cases. For the higher values of Prandtl number (10, 100, 1000), all six laminar cases require use of the Hausen correlation for thermally-developing laminar flow. Twenty-four other cases, all involving either transition (2300 ~< Re < ~10,000) or turbulent flows, use the Gnielinski correlation. All sources recommend the 3.66 value for laminar, fully developed thermally flows, but the other correlations are not necessarily recommended by all authors. Other sources provide correlations for cases not covered by this particular set.

5.2 Internal Flow Correlations

Note that a higher Nusselt number does not necessarily imply a higher heat transfer coefficient than a lower one. The Nusselt number represents the ratio of the conductive resistance (D/k) of a slab whose thickness is equal to the pipe diameter and whose thermal conductivity is equal to that **of the same fluid** to the convective resistance (1/h), or in a more positive tone, the enhancement due to convection over a certain state of conduction only **in that same fluid**. Thus, as seen in Figure 2, a liquid metal (Pr << 1.0, k large) will show a lower Nusselt number than an oil (Pr1>>1.0, k small), but the quantity that is really of interest, the convection coefficient (h), is many orders of magnitude greater for the liquid metal than for an oil. This is a good example of how non-dimensionalization, while often convenient, tends to obfuscate the physics.

Usage

It is well to keep in mind that the convective heat transfer coefficients that this module finds are only part of the story. If the overall calculation were for a pipe passing through a room, the equivalent thermal circuit might be as seen in Figure 3.

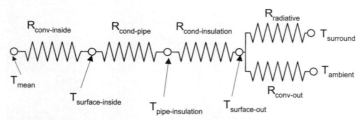

Figure 3. Equivalent circuit for insulated pipe hanging in large room

The correlations discussed in this section and the internal flows module itself are only capable of providing an estimate of the value of the resistor on the far left, that is the convective resistance between the fluid (represented by its mean temperature) and the inside surface temperature of the pipe. The latter would in many cases be an unknown.

Reference

1. Incropera, F.P., and DeWitt, D.P., *Fundamentals of Heat and Mass Transfer*, 4th Ed., Wiley, New York (1996).

6.1 Natural Convection in a Saturated Porous Layer

Introduction

In this module we study the problem of natural convection of fluid contained in the interstices of a porous material. These interstices are assumed to be interconnecting (that is the layer is not just porous, but also *permeable*) and to be completely saturated with the fluid. In one scenario (Figure 1) the lower boundary of the solid matrix is assumed heated, while the top is cooled. The interstitial fluid near the heated bottom, being less dense, tends to rise. At the top it cools and sinks to the bottom, thus setting up a convection cell. In a second configuration (Figure 2) one of the sides is maintained at a temperature higher than that of the other. Such convection could take place within a fibrous insulation, or in an aquifer or geothermal reservoir. There is a similar problem of natural convection within a homogeneous layer of fluid heated from below and cooled at the top; this latter scenario is known as Benard convection. The Benard problem has been solved frequently by numerical means, both because it is interesting in itself and because it makes a good test of numerical schemes.

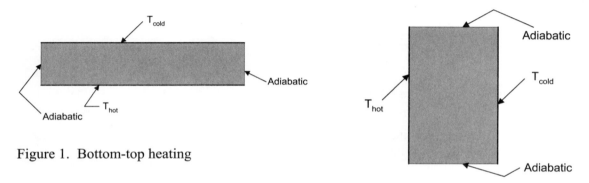

Figure 1. Bottom-top heating

Figure 2. Side-side heating

The governing momentum equations for flow in a porous layer are considerably simpler than those for a homogeneous fluid; this also makes the overall algorithm much simpler to program and faster to execute. Even though this problem is somewhat simpler than Benard convection, it does serve as a good step beyond problems involving only a single partial differential equation. As will be seen in the next section, the governing equations will be reduced to an elliptic equation and a single parabolic (in time) transport equation. Two different algorithms for solving this set of equations, the first an old standby and the other somewhat more modern, are discussed thoroughly in appendices included on the CD-Rom.

Background

In place of the full Navier-Stokes equations, the flow within a porous material is assumed governed by what are known as the Darcy equations. These were derived originally from experimental observations, but it is easy to ascertain their form from consideration of Poiseuille flow in a straight tube (Bear, 1972). Written in component form they are:

$$0 = -\frac{\partial p}{\partial x} - \frac{\mu}{k}u, \tag{1}$$

$$0 = -\frac{\partial p}{\partial y} - \frac{\mu}{k}v - \rho_f g \tag{2}$$

Note that the temporal derivative and advection terms, which appear normally on the left side of the momentum equations, are neglected; that is, the flow is assumed to be "creeping." The Laplacian of velocity terms found in the Navier-Stokes equations reduces to the second term in each of Equations 1 and 2. That is, the viscous resistance, which is felt normally through the "no-slip" condition at solid walls, is modeled as a volumetric resistance that is the result of the no-slip condition applying on a microscale along a multitude of tortuous paths through the material. The k appearing in each equation is the *permeability* of the solid matrix and not to be confused with the thermal conductivity (k_m). It is assumed here to be uniform and isotropic and has dimensions of length squared. Bear (1972) gives typical permeability values of $10^{-3} - 10^{-5}$ cm^2 for clean gravel, $10^{-5} - 10^{-8}$ cm^2 for clean sand and $10^{-9} - 10^{-12}$ cm^2 for stratified clay. The last term in Equation 2 is the body force; this is what causes the flow in the present case. The velocities appearing in these equations are "superficial" velocities; that is they are not the actual velocity in a particular pore, but are averaged over the entire area of a face including both the open area and the solid matrix. Velocities in individual pores would be higher.

With the fluid assumed in thermal equilibrium with the adjacent solid matrix, the governing energy conservation equation is similar to that for a homogeneous, incompressible, low speed flow:

$$(\rho c_p)_f \left[\frac{\partial T}{\partial t} + \frac{\partial uT}{\partial x} + \frac{\partial vT}{\partial y} \right] = k_m \left[\frac{\partial^2 T}{\partial x^2} + \frac{\partial^2 T}{\partial y^2} \right] \qquad (3)$$

The thermal conductivity k_m is that of the fluid-solid mixture, since conduction takes place through both. The heat capacity (ρc_p) multiplying the advection terms is that of the fluid because only the fluid moves. The transient term should really be multiplied by the heat capacity of the mixture; one way of doing that is to introduce a *capacity ratio* to multiply the transient term:

$$\sigma = \frac{\phi(\rho c_p)_f + (1-\phi)(\rho c)_s}{(\rho c_p)_f}$$

(ϕ Is the porosity of the solid), but we will ignore this complication and just use the heat capacity for the fluid. Such an assumption is not that bad. While the densities of typical rocks are 2.5 to three times that of liquid water, their specific heats are correspondingly less.

The continuity equation takes the usual form (except for the velocities being superficial values):

$$\frac{\partial u}{\partial x} + \frac{\partial v}{\partial y} = 0 \qquad (4)$$

The fluid is assumed to be "Boussinesq"; i.e., density changes are everywhere negligible except in the body force term in Equation 2, and there density is taken as a linear function of temperature:

$$\rho_f = \rho_o (1 - \alpha(T - T_c)) \qquad (5)$$

The reference density corresponds to that of the upper boundary, where $T = T_c$.

6.1 Natural Convection in a Saturated Porous Layer

We wish to solve the Equations 1-5 numerically. Fortunately they can be simplified considerably. The pressure may be decomposed into a "motion" pressure and a hydrostatic pressure corresponding to ρ_o; i.e., $p = p^* + p_{hyd}$, where $p_{hyd} = -\rho_o gy$. With this substitution the momentum equations (1,2) become:

$$0 = -\frac{\partial p^*}{\partial x} - \frac{\mu}{k} u, \tag{6}$$

$$0 = -\frac{\partial p^*}{\partial y} + \rho_0 \alpha g (T - T_c) - \frac{\mu}{k} v \tag{7}$$

We also introduce non-dimensional variables as follows:

$$x' = \frac{x}{H}, \quad T' = \frac{T - T_c}{T_h - T_c} = \frac{T - T_c}{\Delta T}, \quad u' = \frac{u}{\kappa_m/H}, \quad t' = \frac{t}{H^2/\kappa_m}, \quad p' = \frac{p^*}{\kappa_m \mu / k} \tag{8}$$

Here the characteristic temperature difference is that from bottom to top, the characteristic length is the thickness H of the porous layer and κ_m is the thermal diffusivity (thermal conductivity divided by density times specific heat) of the fluid-solid mixture. The vertical momentum equation becomes:

$$0 = -\frac{\kappa_m \mu}{kH} \frac{\partial p'}{\partial y'} + \rho_o g \alpha \Delta T \, T' - \frac{\mu \kappa_m}{kH} v' \tag{9}$$

The horizontal momentum equation is non-dimensionalized similarly. Clearing them both up and dropping the primes, one obtains:

$$0 = -\frac{\partial p}{\partial x} - u \tag{10}$$

$$0 = -\frac{\partial p}{\partial y} + Ra\, T - v \tag{11}$$

Here $Ra = \frac{\rho_o g \alpha \Delta T}{\kappa_m} \frac{k}{\mu} H$. This collection of parameters is called the Rayleigh number and measures the importance of the driving force (buoyancy) relative to the restraining force (fluid shear in the interstices). For the base-heated configuration, the Rayleigh number must exceed a certain critical value ($4\pi^2$ for the boundary conditions to be used here) in order for steady natural convection to arise. (The Rayleigh number defined for the more-frequently studied homogeneous flows (Benard convection) has in it the vertical dimension cubed and tends to run orders of magnitude higher than the Darcy-modified Rayleigh number defined here for saturated porous layers.)

6.1 Natural Convection in a Saturated Porous Layer

The energy equation when non-dimensionalized is simply Equation 3 without the coefficients:

$$\frac{\partial T}{\partial t} + \frac{\partial uT}{\partial x} + \frac{\partial vT}{\partial y} = \frac{\partial^2 T}{\partial x^2} + \frac{\partial^2 T}{\partial y^2} \tag{12}$$

The numerical solution of these equations by the older of two approaches (the vorticity-stream function formulation) is discussed in Appendix I; the solution using a somewhat more recent method (primitive variables) is discussed in Appendix II. With only a single parabolic transport equation involving advection terms (Equation 13), natural convection in a porous layer turns out to be a more desirable testbed for evaluating numerical schemes than, say Benard convection, where three transport equations (energy plus horizontal and vertical momentum) involve terms of a similar form. Yet whichever of these two methods is chosen, one deals with a coupled set of partial differential equations.

The Interface and Program Operation

A primitive variable algorithm, the implementation of which is discussed thoroughly in Appendix II on the CD-Rom, has been applied to the governing equations. A second-order upwind scheme was used for the advection terms in Equation 13. The interface itself is seen in Figure 3 below after a run for one particular set of parameters. Unlike other modules in this series, this one has been implemented completely in Visual Basic (Version 6), which unlike VB3 is a compiled language and fast enough for both intensive calculations and graphics.

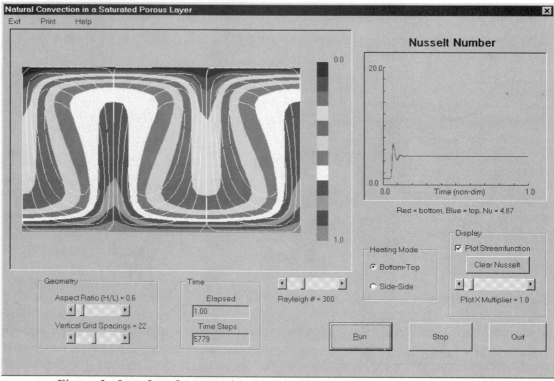

Figure 3. Interface for natural convection in saturated porous layer module

Geometric factors, including the aspect ratio and the number of grid points, may be set at the bottom left. These two parameters are set initially and unlike several others may <u>not</u> be changed during a run. (In fact, these two controls are deliberately disabled during the calculation.) Since we use $\Delta x = \Delta y$, the aspect ratio of the computing region is set by simply adding or subtracting an appropriate number of columns of cells in the horizontal direction. Thus, if one specifies 15 gridpoints in the vertical direction and an aspect ratio of unity, there will also be 15 gridpoints in the horizontal. If one specifies the same number of gridpoints in the vertical, but an aspect ratio (height/length) of 0.1, then there will be 150 gridspacings in the horizontal times 15 in the vertical for a total of 2250 volumes. Similarly, if one specifies a tall, slender region (H/L > 1.0), there will be correspondingly few gridpoints in the horizontal direction.   Since a simple explicit algorithm has been used for the advancement of the single parabolic equation (Eqn. 13), one must be patient if a large number of gridpoints have been specified.

Just to the right of the geometry box is information relating to the time. The non-dimensional elapsed time is reported throughout the transient in the upper box, while the corresponding number of timesteps needed to achieve that time is reported below. The determination of the timestep is taken care of automatically and follows the procedure detailed in Appendix II.

A horizontal slider near the center allows specification of the Rayleigh number. This parameter may be changed during a run, as may the heating direction, for which an option box is provided to the right of that. Here the user may select either bottom-to-top or side-to-side heating. Changing heating mode or Rayleigh number during a calculation produces some interesting transients. Of course, as noted earlier, the bottom-top heating mode has a critical value that must be exceeded before convection will take place. There is no critical value for side-side heating, but at low values, the flow will be quite docile and the isotherms nearly vertical. At the extreme right the user may opt to plot the streamfunction over the isotherms (as seen in the main output window above) and may clear the Nusselt number vs. time plot above at any time. Since there may be some interest in comparing results for different runs, the Nusselt number window is only cleared when the user chooses.

The main output window in the upper left shows isotherms, and if the user chooses, streamlines superimposed on top. This contour plot is made 200 times during any particular transient. For coarse enough meshes (and on a fast enough PC), the plot looks almost "animated."   For the particular Rayleigh number chosen in this example and the aspect ratio selected, a stable, three-roll pattern is quickly set up. Oscillatory patterns may be observed for higher Rayleigh numbers coupled with various aspect ratios. Naturally this calculation produces only 2-D solutions at high values of Rayleigh numbers where 3-D solutions would be observed in nature.

The Nusselt number, the ratio of the actual heat transfer to that were there only conduction, computed along the bottom and the top edges (along the sides for side-side heating) is plotted in the other window. The parameters selected for depiction in Figure 1 result in a relatively docile flow. After an initial transient that lasts less than 0.1 non-dimensional time units, the flow for this aspect ratio and Rayleigh number settles into a stable three cell roll (from left to right, counterclockwise, clockwise and counterclockwise). There are no numerical values given for the streamline increments, but one may ascertain the flow direction from the distortion of the isotherms. One might also note the streamline spacing stays fairly uniform right up to all walls. For a homogeneous flow one would find the spacing widening as one neared a wall because of the decreasing velocity accommodating the no-slip wall.

References

1. Bear, J., *Dynamics of Fluids in Porous Media*, American Elsevier, New York, 1972.

2. Bejan, A., *Convective Heat Transfer*, 2nd Ed., Wiley, New York, Chap. 12, 1995.

3. Gebhart, B., Jaluria, Y., Mahajan, R.L. and Sammakia, B., *Buoyancy-Induced Flows and Transport*, Hemisphere, New York, Chap. 15, 1988.

4. Ribando, R.J. and Torrance, K.E., "Natural Convection in a Porous Medium: Effects of Confinement, Variable Permeability, and Thermal Boundary Conditions," *J. Heat Transfer, Trans. ASME Series C*, 98, pp. 42-48, 1976.

5. Oosthuizen, P.H. and Naylor, D., *Introduction to Convective Heat Transfer Analysis*, WCB McGraw-Hill, Chap. 10, 1999.

6. Nield, D.A., and Bejan, A., *Convection in Porous Media*, 2nd Ed., Springer, NY, 1999.

Porous.doc 7/5/00

7.1 One-Dimensional Heat Exchangers

Introduction

This write-up discusses the first of the two modules developed for the analysis and design of heat exchangers. This one (HX1D) is applicable to those heat exchangers for which the temperature distribution may be approximated as a function of a single coordinate. This group includes concentric-tube heat exchangers and shell-and-tube exchangers (Figures 1, 2). The other algorithm (HX2D) applies to designs such as cross-flow heat exchangers, in which the temperature of one or both fluids must be considered a function of two spatial variables. For one-dimensional heat exchangers, the two governing heat-balances yield two, coupled, *ordinary*-differential equations, while cross-flow heat exchangers are described by two coupled, first-order *partial*-differential equations.

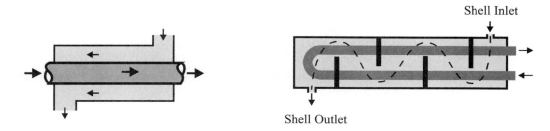

Figure 1. Schematics of sample 1-D heat exchangers: (a) Concentric tube, (b) Shell-and-Tube

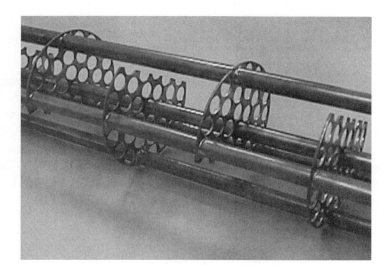

Figure 2. Section from a small shell-and-tube heat exchanger. With the 25 missing tubes in place, the baffles force the shell fluid to take a serpentine path through – as depicted in Figure 1b.

The analytical solutions to similar coupled equations were used in the early 20th century to generate the correction-factor charts used with the Log-Mean-Temperature-Difference (LMTD) approach to heat exchanger *design* [1]. The LMTD method solves a boundary-value problem; the flows (actually heat capacity rates) plus inlet and outlet temperatures are all known, and an iterative scheme is used to find the appropriate "*UA*" product ("*U*" is overall heat-transfer coefficient (e.g., in W/m^{2}K), "*A*" is total heat-transfer area (m^2)) — which in turn governs the sizing of the particular heat exchanger required. Analytical solutions to similar equations lead to the published curves now used with the Effectiveness-NTU (number of transfer units) scheme. This method is normally used to give a measure of *performance* for a given design by solving an initial-value problem. The flows, inlet temperatures and the value of *UA* are known, while the outlet temperatures and heat-transfer rates are the computed results.

While it is clear that a one-dimensional model would be applicable to the concentric tube heat exchanger seen in Figure 1a, it is not so clear that such a model is applicable to the shell-and-tube arrangement of Figures 1b and 2. In the conventional analytical solutions [1] the x-position is defined by the position along the shell. The shell fluid at any position is then assumed to be exchanging heat with fluid at the mean temperature of any tubes at that x location. Thus, for instance, near the right end (x = 0) of Figure 1b, the shell fluid is exchanging heat with tube fluid that has just entered and fluid that is ready to leave the exchanger.

These traditional methods generally yield a few simple parameters characterizing the capability of a particular device. In the slide rule era in which these analytical solutions were developed, it was sufficient to evaluate the "bottom line" parameters for a variety of heat-exchanger configurations and then to make these results readily available in graphical form for the engineering professional. Today virtually all graduate and undergraduate heat-transfer textbooks (e.g., [2,3]) present both the LMTD and the effectiveness-NTU methods — including the many accompanying graphs and descriptive equations — as the primary means of thermal analysis and design for heat exchangers.

In more recent years the popular graduate text of Kays and Crawford [4] suggested the use of electronic spreadsheets to solve a discretized form of the heat-balance equations for heat exchangers — particularly for the analysis of cross-flow exchangers. Additionally, the power-plant design textbook of Li and Priddy [5] presents a detailed solution for an evaporative cooling tower, where — because of the phase change and the accompanying property variations — a numerical solution is not only desirable but, in fact, necessary. Our alternative approach [6] extends these suggested implementations to a more comprehensive numerical methodology.

In contrast to the traditional analytical methods, the numerical/graphical approach we present here offers enhanced insight by allowing the user to visualize the entire temperature distribution for both fluids in the heat exchanger. With the complete temperature distribution — plus all the conventional performance measures — immediately available on a display monitor, the user can quickly study and understand the effects of various physical parameters. Perhaps most importantly, students are introduced to heat-exchanger thermal-design as a straightforward application of fundamental physical principles — in this case, the conservation of energy noted earlier.

Complete details of the numerical procedures used in this module are provided in an appendix found on the CD.

HX1D – The Interface and Program Operation

A sample of the user-interface for this module is presented in Figure 3. The upper-right quadrant of the user interface is available for user input of the problem definition and for selection of the method of calculation; the white boxes are used for text and numerical input. Heat-capacity rates $(\dot{m}c_p)$ for the two fluids are specified directly in appropriate units, typically W/K. The selection of calculation method requires an associated input of desired shell outlet-temperature (for a design calculation) or of *UA* (for a performance calculation). The choice of shell outlet temperature as the iteration variable for the former was quite arbitrary and would not be appropriate, for instance, were that fluid undergoing a phase change. For such cases one can use the performance option and iterate manually until the desired conditions are reached.

7.1 One-Dimensional Heat Exchangers

User-definition of unique nodal-numbering arrangements are provided in separate windows, which are brought into view by clicking on the "Modify" button located to the right. The special user-input window for automatic numbering and display of certain shell-and-tube exchangers is depicted in Figure 4. This input form assumes a configuration based on two tube-passes per shell-pass. The user can select the number of shell passes (up to six) and the number of baffles per shell. Pressing the "Update" button initiates the computation of a new configuration and its display. The nodal numbering system itself is somewhat complex, and is unique to this particular numerical approach — thus its computation is performed automatically without a lot of unnecessary user involvement.

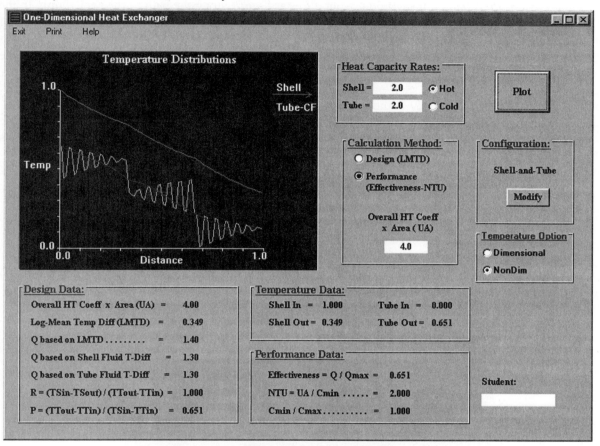

Figure 3. User interface for 1-D heat exchanger analysis

The sample input-window shown in Figure 4 corresponds to the three-shell-pass geometry associated with the display of Figure 3. The particular selection shown is for three shell-passes with two tube-passes/shell-pass and 12 baffles/shell – giving a total of 78 control volumes. (As noted earlier, we have found it convenient here to use the baffles as boundaries for the small control-volumes, although this is not a requirement.) A second user-input window is provided for numbering the nodes of annular-flow and twin-tube exchangers, both parallel and counterflow. In this case, the user simply selects the flow direction and the number of nodes. Numbering of these nodes is also done automatically, even though the process is trivial by comparison with the shell-and-tube system. A third user-input window allows the user to input connectivity data manually for any 1-D geometry not covered adequately by the other two input options. If one wants to match the corresponding analytical solutions (those the LMTD correction and effectiveness-NTU charts are based on), then one is advised to specify the maximum number of baffles allowed. The analytical solutions are in effect based on an infinite number of baffles – corresponding to perfect mixing at a particular x location.

7.1 One-Dimensional Heat Exchangers

In the Temperature Option panel to the extreme right, the user can select whether temperatures are to be non-dimensionalized by the approach temperature difference (the default) or are to be input and returned in dimensional form. In the latter case textboxes are made visible for data entry.

When the user selects the "Plot" button, the numerical algorithm is initiated using the input data specified by the user. The plot of temperatures as a function of position has some interesting features. In Figure 4, the temperature profile is shown in the traditional "counterflow" manner, i.e., hot shell-fluid running from left-to-right and cold tube-fluid running from right-to-left. To this, we include an additional profile that shows the computed tube-fluid temperatures as "seen" along the path of the shell-fluid. In the particular geometry computed here (and shown in Figure 5), the hot shell-fluid entering the first shell sees first the exiting tube-fluid and then encounters the colder tube-fluid—itself just entering the shell. The shell-fluid then interacts with tube-fluid almost ready to leave, followed by tube-fluid that is only slightly heated, and so on ... Upon entering the second shell, the shell-fluid encounters the tube in the middle of its run, and here the "wiggles" seen along the shell-path begin small and increase in amplitude as the path proceeds through the second shell. Similarly in the third shell, the shell-fluid enters where the difference in temperature is large between tube-legs and leaves midway along the tube pass where it is minimal. The seemingly odd behavior depicted in the plots thus helps to promote a new insight into the internal physics of heat-exchanger operation.

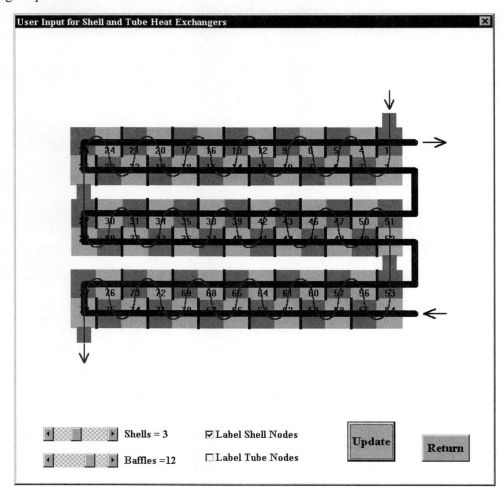

Figure 4. Sample user-input window for shell-and-tube configuration

7.1 One-Dimensional Heat Exchangers

Computed numerical data is provided in the lower third of the screen. Parameters associated with the conventional LMTD method are reported in the lower left. These include the UA product (an output for a design calculation, but an input for a performance run) and the LMTD. Three separate values for the heat transferred are reported. One Q is based on the temperature drop of the hot fluid, while the second is based on the temperature rise of the cold fluid. At convergence these should match closely. The third Q is based on the value of UA (which is input for performance calculations and output for design) and the LMTD. Since this value has not been "corrected," then when divided by either of the other computed Q's one has found the LMTD correction factor normally taken from charts. If the non-dimensional temperature option has been selected, then all reported temperatures and all reported heat transfers must be multiplied by the approach temperature difference (T_{hot-in}-$T_{cold-in}$) to obtain appropriate dimensional values. The R and P ratios facilitate checking these results with the LMTD correction charts. For cases involving a phase change (evaporators and condensers) and thus no appreciable temperature change, the appropriate Q values are dimmed to indicate their lack of reliability. Parameters associated with the effectiveness-NTU method are reported in the bottom center.

A number of safeguards have been built into the program to help eliminate data inputs that would hinder or prevent convergence of the solutions. These checks represent necessary, but not necessarily sufficient conditions. In many cases the cause for non-convergence may be readily found by looking at the output graphs. For design calculations, where the desired value of *UA* is sought, a check is performed to determine whether the requested outlet temperature would result in effectiveness greater than one; if so, the user is requested to supply a different specified outlet temperature. For performance calculations, input values of *UA* that would yield very large values of NTU are automatically rejected. (The conventional effectiveness charts typically stop at NTU values of 5.0.) Very large values are of no practical worth because the configurations they represent would have all the heat being transferred in only a tiny fraction of the overall area—a fraction right near the shell inlet.

References

1. Bowman, R.A., Mueller, A.C., and Nagle, W.M., "Mean Temperature Difference in Design," *Transactions of the ASME*, Vol. 62, pp. 283-294.

2. F.P. Incropera and D.P. DeWitt, *Fundamentals of Heat and Mass Transfer*, 4th Ed., Wiley, New York, 1996, pp. 581-632.

3. A.F. Mills, *Heat and Mass Transfer*, Irwin, Chicago, 1995.

4. W.M Kays and M.E. Crawford, *Convective Heat and Mass Transfer*, 3rd Ed., McGraw-Hill, New York, 1993, pp. 417-442.

5. K.M. Li and A.P. Priddy, *Power Plant System Design*, Wiley, New York, 1985, pp. 292-298.

6. Ribando, R.J., O'Leary, G.W, and Carlson, S.E., "A General, Numerical Scheme for Heat Exchanger Thermal Analysis and Design," *Computer Applications in Engineering Education*, Vol. 5, pp. 231-242, 1997.

7.2 Two-Dimensional Heat Exchangers

Introduction

This write-up discusses the second of the two modules developed for the analysis and design of heat exchangers. This algorithm (HX2D) applies to designs, including cross-flow heat exchangers, in which the temperature of one or both fluids must be considered a function of two spatial variables. Both heat exchanger modules involve a numerical solution of the two coupled, governing heat-balance equations in discretized form. While one-dimensional heat exchangers (shell-and-tube, etc.) are described by two coupled ordinary-differential equations, the cross-flow heat exchangers covered by this module are described by two coupled, first-order partial-differential equations. That mathematical disparity requires different algorithms and different displays of the data, and thus we chose to create two separate modules.

Using complicated integral methods, Nusselt [1] solved those partial-differential equations analytically, and his results now form the basis for detailed graphs that accompany most textbook coverage of these particular heat exchangers. Unfortunately, the graphs and the associated methods do tend to absorb the user in a cloud of routinized problem solving—without offering much opportunity to consider the actual processes of thermal exchange within the hardware itself. As an alternative, we have implemented a numerical approach for solving the coupled first-order equations of these 2-D exchangers using an iterative technique quite similar to that described earlier for 1-D heat exchangers.

The popular graduate text of Kays and Crawford [2] suggested the use of electronic spreadsheets to solve a discretized form of the heat-balance equations for heat exchangers—particularly for the analysis of cross-flow exchangers. Additionally, the power-plant design textbook of Li and Priddy [3] presents a detailed solution for an evaporative cooling tower, where—because of the phase change and the accompanying property variations—a numerical solution is not only desirable but, in fact, necessary. Our alternative approach presented here extends these suggested implementations to a more comprehensive numerical methodology.

The numerical algorithm we use for the typical single-pass, cross-flow heat exchangers shown in Figure 1 employs an approach similar to that used for shell-and-tube heat exchangers. As previously noted, in shell-and-tube exchangers the temperature in each fluid is considered to be a function of only one spatial coordinate; thus resulting in governing equations specified as ordinary-differential equations. For a cross-flow heat exchanger, the temperature distribution in one or both of the fluids is a function of two directions (although the flows themselves are both considered uni-directional); thus resulting in governing relations expressed as partial-differential equations. The details of the algorithm are covered thoroughly in the appendix.

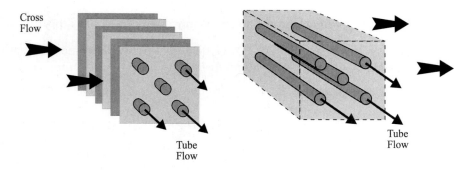

Figure 1. Schematics of typical 2-D heat exchangers: (a) Unmixed, (b) Mixed

7.2 Two-Dimensional Heat Exchangers

HX2D - The Interface and Program Operation

Figure 2 presents the user-interface along with some typical input (the white boxes) and computed results. The mid-section of this interface is available for user input of problem definition and the method of calculation. Since no geometry-specific nodalization is needed here, the input for the single-pass crossflow (2-D) exchanger is easier than that for 1-D exchangers. One must select the configuration from the five options provided in the bottom right corner. The first four (neither fluid mixed, one or the other mixed and both mixed) are options for which LMTD correction charts and effectiveness-NTU charts are commonly provided in textbooks. The fifth is for a specific two-pass geometry to be discussed later. As with the 1-D module, the user must provide the heat capacity rates for the two fluids in consistent units. Condensers and evaporators can be modeled by specifying one or the other heat capacity as being very much bigger than the other.

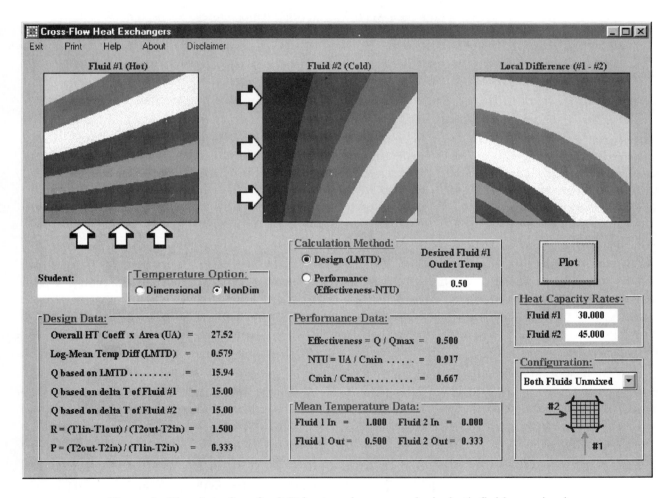

Figure 2. User Interface for 2-D heat exchanger analysis, both fluids unmixed

If, under Calculation Method, the user selects the Performance (Effectiveness- NTU) option, then they must provide the UA value (overall heat transfer coefficient times area in the same units as the heat capacity rates) and no iteration is required. If instead the Design (LMTD) option is used, then they must provide the desired outlet temperature for the hot fluid, and an iterative calculation proceeds until a UA value that yields that temperature is found. It is very easy to specify input values for which there is no possible solution, and such cases are flagged. The outlet temperature of the hot fluid was selected quite arbitrarily as the iteration variable. Clearly that value is inappropriate for a condenser, where that temperature is not intended to change. At other times the total heat transfer for the exchanger may be the design goal. In these later cases one can simply select the Performance option and iterate manually until the desired values are reached. The "dimensional" temperature option allows just that; inlet and outlet temperatures plus the desired outlet temperature for the Design option can be input in temperature units consistent with other inputs.

As can be seen in Figure 2, the numerical data provided as output are basically the same as for the 1-D module. Similarly, selecting the "Plot" button initiates the numerical computations that are performed by Fortran routines contained in a separate Windows DLL. Even though a 30x30 grid is used, the results are returned nearly instantly. Numerical data most closely associated with the conventional LMTD method are displayed in the bottom left panel. The UA product may be either an input (for Performance calculations) or the output (for design). The log mean temperature difference (LMTD) is reported in the next line. With the non-dimensional temperature option, the LMTD like all other temperatures, has been non-dimensionalized by the approach temperature difference ($T_{hi}-T_{ci}$). Below the LMTD are values of the overall heat transfer computed three different ways. The first comes from $Q = UA\,\Delta T_{lm}$, the second from the temperature drop of the hot fluid ($Q = (\dot{m}c_p)_1 \Delta T_1$), while the third comes from the temperature rise of the cold fluid ($Q = (\dot{m}c_p)_2 \Delta T_2$). If the particular heat exchanger involves a correction to counterflow, then the first of these, the one computed from the LMTD, must be corrected using textbook charts before it will match the other two. In fact, the ratio of the first value to the second or third may be taken as a measure of the deviation from the ideal counterflow geometry and a measure of the quality of the design. If the input parameters indicate a condenser or evaporator (one heat capacity rate much greater than the other), then the appropriate computed Q will be dimmed to indicate that, since based on a nearly-zero temperature change, it is probably not reliable. With the non-dimensional temperature option, all three values of Q must be multiplied by the approach temperature difference. The last two values returned in the lower left (R and P) are the parameters used in the conventional LMTD correction charts and are presented merely to facilitate comparison with textbook values. For an evaporator the value of R will be dimmed.

Parameters associated with the effectiveness-NTU method are reported in the middle center. These include the computed effectiveness (Q/Q_{max}), the Number of Transfer Units (NTU) and the capacity ratio (C_{min}/C_{max}). These values are presented to facilitate comparison with charts. Finally at the bottom center values of the inlet and outlet temperatures (divided by the approach temperature difference for the non-dimensional temperature option) for both fluids are given.

In the present case, since one or both fluid temperatures are functions of two spatial-coordinates, we present color-contour plots displaying the temperature distributions for each fluid. A third contour plot is included to display the local difference between the two fluid temperatures. This latter plot provides useful insight as to the general quality of a particular exchanger design. It is particularly interesting to note that when the computed value of effectiveness (q/q_{max}) is very high, then the temperature difference over much of the exchanger is generally low—resulting in considerable wasted capacity.

It can be difficult to visualize how these color contour plots relate to the cross-flow heat exchangers shown in schematic form in Figure 1. Figure 3 shows another cross-flow configuration, a single-pass, plate-fin heat exchanger. Here the fins, shown as the short vertical lines, prevent each fluid from mixing laterally. If we were looking down and took a hypothetical slice through the second, fourth or sixth layer from the bottom, then the temperature field we would see corresponds to that shown for the hot (#1) fluid in the contour plot. Similarly a slice through the first, third or fifth would correspond to the middle contour plot for the cold (#2) fluid. If the fins were removed in either or both of the hot and cold channels, we would have one of the three "mixed" configurations allowed by the module. In trying to reconcile the schematics of Figure 1 with the corresponding contour plots, one must envision many more tubes than

7.2 Two-Dimensional Heat Exchangers

shown in those depictions and that they are "homogenized." Clearly one would also be looking down from above.

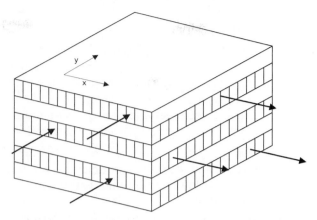

Figure 3. Single-Pass Cross-Flow Exchanger

The final configuration option corresponds to the automotive intercooler shown in Figure 4 below. Air which has been heated in the turbocharger enters at the top right, travels from right to left through the five upper channels, mixes in the plenum on the left and then flows left to right through the bottom five channels and on to the engine. Both the cooling air channels, which are seen in this photo, and the hot air channels (which may be seen by looking up the outlet) are finned. From a programming standpoint, the addition of this configuration was trivial. Both fluids are considered unmixed (because of the fins), but halfway through its traverse, the hot fluid is mixed within itself. (The mixing procedure is discussed in the appendix.) In viewing the resulting contour plots, one must imagine oneself looking down from above, with the heat exchanger being split halfway down (between the passes) and "unfolded." The "kinks" seen halfway along in all three contour plots may then be explained.

Figure 4. Automotive intercooler

References

1. M. Jakob, *Heat Transfer* - Vol. II, Wiley, New York, 1957, pp. 230-245, 217-227.

2. W.M. Kays and M.E. Crawford, *Convective Heat and Mass Transfer*, 3rd Ed., McGraw-Hill, New York, 1993, pp. 417-442.

3. K.M. Li and A.P. Priddy, *Power Plant System Design*, Wiley, New York, 1985, pp. 292-298.

8.1 Transmissivity of Glass

Introduction

In this project we investigate the radiative properties of two particular types of glass: one of them a standard glass and the other what is called a "low – E" (for emissivity) glass. Stop by any glass shop and you can pick up literature on various brands of the latter. You will find that many modern windows are made of two sheets of glass, at least one of them low-E, and with, perhaps, argon or krypton in the gap between them. In such glass manufactured for residential application, the special coating is not visible to the eye, but its presence may be determined with an ohmmeter since it is electrically conductive. A data file of the spectral transmissivity (τ_λ) is supplied for two representative glasses, and we will determine the total transmissivity of each glass for both solar (short wavelength) and terrestrial (long wavelength) radiation.

Recall that unlike total emissivity, which is a function of the temperature of the surface itself, the total transmissivity (and total absorptivity and reflectivity) are functions of the <u>irradiation</u>. That is, the integration of the spectral value to get the total value, which we will do both using the conventional, tabulated blackbody radiation functions and also numerically, is done over the spectrum corresponding to the temperature of the source, which here will be taken as a blackbody. Therefore, for both glasses you will find quite different values of total transmissivity depending on whether the incoming radiation is high temperature solar (short wavelength) or low temperature terrestrial (long wavelength) in origin. This significant difference in transmissivity to short and long wavelength radiation leads to what is known as the "greenhouse effect," a term which more recently has been extended to the analogous effect in the atmosphere caused by the accumulation of CO_2 and water vapor.

The Problem

Data for the transmissivity of two glasses, standard and low-emissivity, as a function of wavelength is given in the second sheet of the Planck's Law workbook found on the CD and is plotted on the next page. Both regular and low-E glass has high transmissivity (.8 - .9) in the visible portion of the spectrum (.4 - .7 μm). The transmissivity of the low-E glass is seen to drop much faster than that of ordinary glass in the near infrared region, meaning that with it the user still gets the visible light while reducing the overall solar gain. It is not obvious where the term "low-E" comes from, but since $\alpha + \rho + \tau = 1$, low transmissivity and high reflectivity imply low absorptivity, which means low emissivity, since $\alpha_\lambda = \varepsilon_\lambda$. "Low-E" thus refers strictly to the behavior of the glass in the infrared (long wavelength) part of the spectrum. (If one were doing a radiative heat transfer analysis for the interior of a room, the window would be considered an opaque surface with surface resistance = $\dfrac{\rho_i}{A_i \varepsilon_i}$. Then high reflectivity and low emissivity implies a high resistance to radiative heat transfer.)

In your calculations assume the sun is a blackbody source at 5800K and produces an irradiation of 1100 W/m^2 on the glass - which might be the windshield of your car. (The irradiation value includes the effect of attenuation in the earth's atmosphere that reduces it from the value of the solar constant (1353 W/m^2)). The solar constant corresponds to that which arrives per square meter at the orbital radius of the earth from a blackbody source having the

8.1 Transmissivity of Glass

diameter of the sun at the sun's apparent temperature. The terrestrial source is assumed to be black and at 300K.

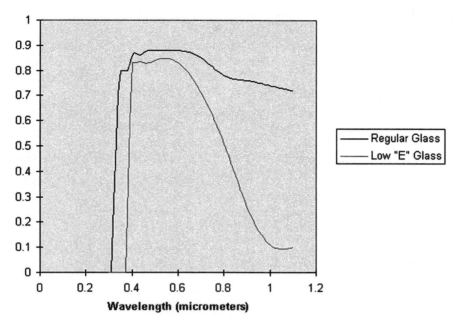

Figure 1. Spectral transmissivity of regular and "Low E" glass

You are to do the following and comment on your findings:

1. Access the Planck's Law workbook. One of the two worksheets includes a table of the transmissivity data for both glasses as seen in graphical form above.

2. Approximate both transmissivities as a constant over a particular wavelength interval (make a good "eyeball" estimate based on the plots) and 0.0 outside that range, i.e., as a "hat" function. Sketch your function for both glasses in your notes.

3. Using the blackbody radiation function table in the text with the "hat" functions you have chosen, estimate the total transmissivity of both glasses to irradiation from the sun. Assume the sun is a blackbody at 5800K.

4. Do this calculation again to estimate the total transmissivity of both glasses to the irradiation from a terrestrial source. Assume a blackbody at 300K. A table for recording your results is given on the next page.

5. Now use Simpson's Rule as described in the Appendix to do a numerical integration and thus get better estimates for all four quantities. You should now have eight total transmissivities:

8.1 Transmissivity of Glass

	Regular Glass B.B.Rad. Func.	Regular Glass Simpson's Rule	"Low E" Glass B.B.Rad. Func.	"Low E" Glass Simpson's Rule
Solar Radiation (5800K)				
Terrestrial Radiation (300K)				

Computed Transmissivities

6. Using the transmissivity values found in Step 5 (which presumably are more accurate than those found in Step 4), find the total solar energy transmitted through both glasses (W/m^2). Use the value of 1100 W/m^2 given above for solar irradiation.

7. Again using your values from Step 5, find the total energy transmitted from the terrestrial source (blackbody at 300K) for both glasses (W/m^2). Assume the source and the glass are two closely spaced parallel plates.

	Regular Glass	"Low E"
Solar Radiation		
Terrestrial Radiation		

Transmitted Radiation (W/m^2)

8. Speculate on how you might profitably incorporate low-E glass into the design of a home. Consider both summer and winter scenarios. Include sketches.

References

1. *1997 ASHRAE Handbook - Fundamentals*, American Society of Heating, Refrigerating and Air-Conditioning Engineers, Inc. Atlanta (1997).

Note: Some sources, including the ASHRAE Handbook, make a distinction between the "ivity" ending used here, and the "ance" ending. Technically the transmissivity, reflectivity, etc. refer to inherent properties of a bulk sample of the material, while the "ance," e.g., transmittance, refers to the property of a specific sample or thickness of a substance or combination of substances.

2. Çengel, Y.A., *Heat Transfer – A Practical Approach*, WCB McGraw-Hill, 1998, pp. 752-759.

3. Chapra, S.C., and Canale, R.P., *Numerical Methods for Engineers*, 2nd Ed., McGraw-Hill, New York (1988), pp 490-494.

8.1 Transmissivity of Glass

Appendix - The Numerical Integration

Problems like that given here are discussed in all heat transfer textbooks, but in those problems the variation of the transmissivity with wavelength is taken to be a very simple "hat" function - e.g., a constant over a particular wavelength range and identically 0.0 or another constant over the rest of the spectrum. Evaluating total properties can then be done easily using the blackbody radiation function table in the text. In order to force you to do a numerical integral of the following definition of the total transmissivity,

$$\tau = \frac{\int_0^\infty \tau_\lambda(\lambda) G_\lambda(\lambda) d\lambda}{\int_0^\infty G_\lambda(\lambda) d\lambda} \quad (1)$$

we have given you more detailed transmissivity data at a large number of equally spaced points. Assume zero transmissivity outside the range of the given data. Since the irradiation is assumed to come from a blackbody source, then,

$$G_\lambda(\lambda) = E_{\lambda,B}(\lambda,T) = \frac{C_1}{\lambda^5 \left[e^{\frac{C_2}{\lambda T}} - 1 \right]} \quad (2)$$

Here $C_1 = 3.742 \times 10^8 \; W \cdot \mu m^4 / m^2$ and $C_2 = 1.439 \times 10^4 \; \mu m \cdot K$. The Planck's Law workbook has a VBA function already prepared for this expression and you can use it in your calculations exactly as if it were a supplied function like the sine or cosine.

You will probably want to use Simpson's 1/3 rule for doing the integration:

$$\int_a^b f(x) dx \approx \frac{h}{3} \left[y_0 + 4y_1 + 2y_2 + 4y_3 + 2y_4 + \ldots + 2y_{n-2} + 4y_{n-1} + y_n \right] \quad (3)$$

Here $y_0, y_1 \ldots y_n$ are the ordinates of the curve $y = f(x)$ at the x values $x_0, x_1 = a + h, \ldots x_n = a + nh = b$. Alternatively this may be written:

$$\int_a^b f(x) dx \approx (b-a) \frac{f(x_o) + 4.0 \sum_{i=1,3,5}^{n-1} f(x_i) + 2.0 \sum_{j=2,4,6}^{n-2} f(x_j) + f(x_n)}{3n} \quad (4)$$

For the given data file $n = 80$, i.e., there are 81 data points. (You need an even number of intervals to use Simpson's 1/3 Rule.) Note that the integral in the denominator of Equation 1 above should come out to σT^4, which suggests a way to check your implementation of Simpson's Rule (realizing that you do have to cut it off somewhere short of infinity) using the table of blackbody radiation functions given in the text.

Transmiss.doc 6/19/00

8.2 Calculation of Radiation View Factors by Nusselt Unit Sphere Method

Introduction

In this exercise we use a modern numerical algorithm to compute directly the view factor between arbitrarily oriented rectangular plates. The view factor between two surfaces is the fraction of radiation leaving the first surface that impinges directly on the second. The view factor from surface i to surface j is given by [1]:

$$F_{ij} = \frac{1}{A_i} \int_{A_i} \int_{A_j} \frac{\cos\vartheta_i \, \cos\vartheta_j}{\pi R^2} dA_i \, dA_j$$

Here ϑ_i is the angle between the surface normal to dA_i and the vector R connecting it to dA_j. This quadruple integral can be done analytically for a few simple geometries; the results may be given as equations or in the form of view factor charts [1-3]. Other means, which are often lumped in the category of "shape factor algebra" and include the reciprocity rule, summation rule and intuition, may be used to extend the utility of the analytically-based results.

In recent years a number of critical applications of radiative heat transfer have necessitated the determination of view factors among thousands and even tens of thousands of surface pairs - a job definitely not to be done by hand! One interesting application is the International Space Station. Since in the vacuum of space, radiative heat transfer is the only mode of heat transfer, the thermal analysis of a complicated space structure may require the computation of thousands of pairs of view factors [4]. Similarly in computer graphics, the radiosity method, which is identical to what thermal engineers use for radiative heat transfer analysis of diffusively emitting and absorbing surfaces, has been adopted widely over the last decade or so for the rendering of complex three-dimensional scenes [5, 6]. Of course, while we thermal analysts include the entire thermal spectrum (ultraviolet, visible and infrared) in our calculations, in computer graphics it is only the visible part of the spectrum that is generally of interest.

A number of numerical techniques for evaluating the above integral have been developed over the past several decades [7]. Among them are the Monte Carlo Ray Tracing Method and double area integration. The latter may be implemented in a straight-forward manner [8]; each plate is subdivided into a large number of elements and a quadruple "Do Loop" is set up to evaluate and sum the terms. The same calculation may be implemented using finite-element techniques [9]. Both methods suffer in accuracy when the differential elements are relatively close together and/or when one dimension of a surface is much longer than the other.

The algorithm we used here is a computational implementation of the Nusselt Unit Sphere method, which was first proposed as an experimental means of determining view factors for odd geometries (2, 10 - 12). Our implementation follows Reference 12 very closely. In this algorithm we "discretize" only the emitting plate (we use 30 subdivisions in each direction) and center an imaginary hemisphere of unit radius in turn over each small area (designated as dA_1 in Figure 1). Then we draw rays from the current origin to each vertex (four in our case) of the receiving surface (Surface 2). The intersections of the four rays with the hemisphere are then found. (The projections of the straight edges of the receiving plate on the hemisphere will be "great circles" and that projection is designated as dA_2'.) A second projection of the figure generated by the projection of the receiving surface on the hemisphere, this time to the base of the hemisphere, yields the view factor as the fraction of the base circle covered by the second projection (dA_2''). Though it sounds complicated, the algorithm takes only about 40 lines; it and the viewing algorithm used to create the display seen on the main form were assigned as projects in an undergraduate computer graphics class several times.

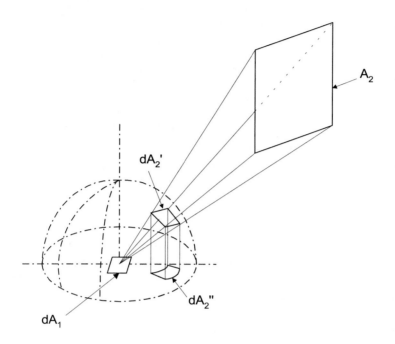

Figure 1. Schematic for Nusselt unit sphere method (redrawn from Reference 12)

Program Operation

This application consists of a main form and two input forms, one for each of the two plates. Upon clicking on **Modify Red Coordinates** on the main form, the input screen for the red (originating) plate appears. Note that this form only allows the user to input the X, Y, and Z coordinates for three of the vertices (in the white boxes seen in Figure 2). Then, under the assumption that the plate is a parallelogram, the program computes the fourth itself (whenever you click the Update button). This ensures that the points will be coplanar. The coordinates must be entered such that in going from Vertex 1 to Vertex 2 to Vertex 3, the surface in question, and not its backside, are to the right if you were to walk along that path. (See Figure 2.) This is a standard convention used in computer graphics and is necessary in order to compute the surface normal properly. Inputting coordinates such that the surface normal faces away from the other plate (rather than towards) is the most common error encountered in using this program. Handy little "manipulatives" such as seen in Figure 3 and provided in the Appendix will help you with this task. Unless you click the **Update** button before clicking the **Close** button, your entries will not register. A similar form is used to input coordinates for the blue (receiving) plate and the same rules hold.

8.2 Calculation of Radiation View Factors by Nusselt Unit Sphere Method

The plates you see on the red plate and blue plate input forms are merely schematic representations and will not change with your data. The scaleqd, perspective drawing of the two plates will only appear on the main form. All of the red surface must be able to "see" all of the blue surface and vice versa. If this condition is not satisfied, a warning will be returned. A full-blown commercial version of this algorithm would subdivide the plates as necessary when there is partial shading or obstruction by another surface, but that feature is not included here.

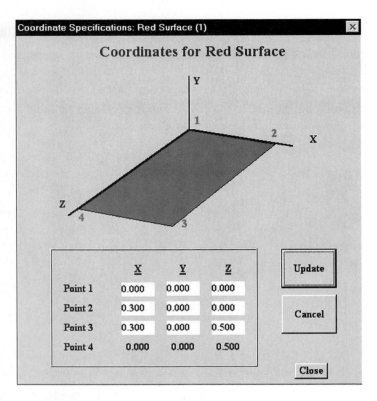

Figure 2. Input screen for red (emitting) plate

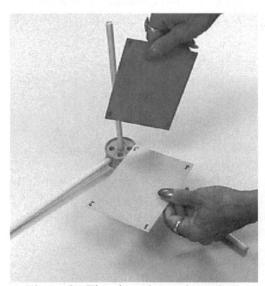

Figure 3. Figuring plate orientations

Once you click the **Close** button on either input form, the main form (Figure 4) will show you a perspective drawing of the plates on the screen and automatically compute the view factor for the current configuration. If you still have to input the second plate, you can ignore this value and any warnings. You can use the slider bars at the upper right to look at the configuration you have set up from other viewpoints. Oftentimes you can use this display to ferret out input data errors.

8.2 Calculation of Radiation View Factors by Nusselt Unit Sphere Method

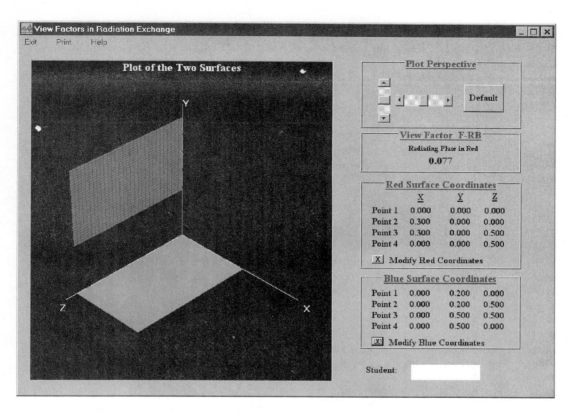

Figure 4. Main form showing plates, coordinate data and computed view factor

Verification

View factor values found using the conventional means, e.g., intuition, charts, analytically derived equations, etc., may and have been used to test this algorithm [13]. An Excel/VBA workbook generated specifically to verify this numerical algorithm is discussed in the next section. For geometries not covered by such means, a few necessary, though not sufficient, means are available. These include engineering judgment, the summation rule and reciprocity checks.

References

1. Incropera, F.P., and DeWitt, D.P., *Fundamentals of Heat and Mass Transfer*, 4th Ed., Wiley, New York, 1996.

2. Modest, M.F., *Radiative Heat Transfer*, McGraw-Hill, New York, 1993.

3. Howell, J.R., *A Catalog of Radiation Configuration Factors*, McGraw-Hill, New York, 1982.

4. Chin, J.H., Panczak, T.D. and Fried, L., "Spacecraft Thermal Modeling," *International Journal for Numerical Methods in Engineering*, Vol. 35, No. 4, 1992, pp. 641-653.

5. Goral, C.M., Torrance, K.E., Greenberg, D.P., and Battaile, B., "Modeling the Interaction of Light Between Diffuse Surfaces," *Computer Graphics*, Vol. 18, No. 3, pp. 213-222, 1984.

6. Hearn, D. and Baker, M.P., *Computer Graphics*, 2nd Ed., Prentice Hall, Englewood Cliffs, NJ, 1994.

8.2 Calculation of Radiation View Factors by Nusselt Unit Sphere Method

7. Emery, A.F., Johansson, O., Lobo, M., and Abrous, A., "A Comparative Study of Methods for Computing the Diffuse Radiation View factors for Complex Structures," *Journal of Heat Transfer,* Vol. 113, 1991, pp. 413-422.

8 Ribando, R.J., and Shi, Q., "Fortran 90 and the Direct Calculation of Radiation View Factors," *Computer Applications in Engineering Education*, Vol. 3, No. 2, pp. 133-137, 1995.

9. Chung, T.J., "Integral and Integro-Differential Systems," *Handbook of Numerical Heat Transfer*, Minkowycz, W.J., Sparrow, E.M., Schneider, G.E., and Pletcher, R.H., editors, Wiley-Interscience, New York, 1988.

10. Farrell, R., "Determination of View Factors of Irregular Shape," *ASME Journal of Heat Transfer,* Vol. 98, no. 2, pp. 311-313, 1976.

11. Lipps, F.W., "Geometric Configuration Factors for Polygonal Zones Using Nusselt's Unit Sphere," *Solar Energy*, Vol. 30, No. 5, 1983, pp. 413-419.

12. LEMSCO, *Thermal Radiation Analyzer System (TRASYS) User's Manual*, Lockheed Engineering and Management Services Company Report LEMSCO-22541, Prepared for NASA LBJ Space Center, April 1988.

13. Ribando, R.J. and Weller, E.A., "The Verification of an Analytical Solution - An Important Engineering Lesson," *Journal of Engineering Education*, Vol. 88, No. 3, July 1999, pp. 281-283.

Acknowledgment

We are indebted to Dana Gould of NASA Langley Research Center for bringing the TRASYS implementation of the Nusselt sphere method to our attention.

Appendix – Sample Manipulative for Getting the Vertex Numbering Correct

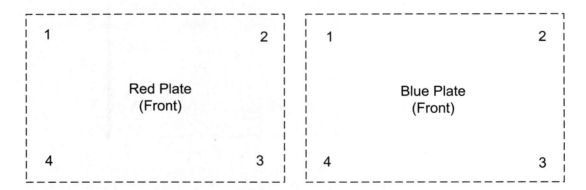

Figure 5. Handy manipulatives for ordering vertices of plates. Cut them out and color them!

8.3 View Factor Spreadsheets and Charts

Spreadsheets

A workbook that calculates the viewfactors for several configurations having analytical solutions is provided on the CD. Charts displaying solutions for three of these geometries are a staple of nearly all heat transfer texts [1], while the fourth is sometimes included [2,3]. View factors for two of these four configurations may also be computed using the Nusselt Unit Sphere module; indeed, one of these worksheets was created during the development of that module specifically to aid in its verification [4]. The third and fourth configurations involve geometries, disks and cylinders, for which our Nusselt module is not applicable. All four worksheets determine the numerical value of the viewfactor and, in addition, draw the configuration to scale on the screen. The latter feature aids greatly in ensuring that the input parameters are indeed those that the user intended. The programming and graphics procedures used are discussed in the Excel/VBA primer contained on the accompanying CD.

The worksheet for perpendicular rectangles having a common edge is seen in Figure 1. Inputs are the dimension of the common edge and the other dimension of the two rectangles. The input boxes are color coded with the figure. A VBA function evaluates the viewfactor using the (somewhat complicated) analytical solution and returns that numerical value. Another VBA function draws an isometric view of the two rectangles on the screen. Running a few sets of parameters helps develop a sense for the magnitude of the viewfactor. For instance if one specifies a 1.0 x 1.0 base and then increases the other dimension of the receiving plate, the viewfactor increases to an asymptotic value of 0.25. Then as the length of the common edge is increased, the viewfactor increases to an asymptotic value of 0.50.

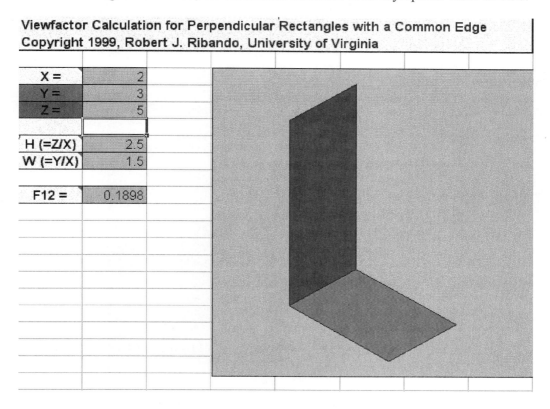

Figure 1. Worksheet for perpendicular rectangles with a common edge

8.3 View Factor Spreadsheets and Charts

Another Excel/VBA worksheet for coaxial parallel disks is seen in Figure 2 below. Inputs are the radii of the two disks and their separation distance. This module returns the numerical value for the viewfactor and allows the user to view the configuration from different angles.

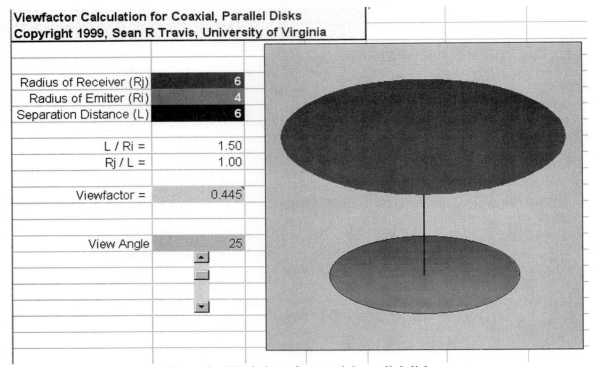

Figure 2. Worksheet for coaxial parallel disks

Still another one for finite, coaxial cylinders [5] is shown in Figure 3.

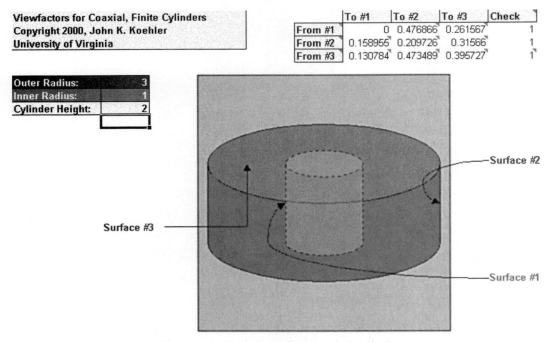

Figure 3. Worksheet for coaxial cylinders

8.3 View Factor Spreadsheets and Charts

Charts

High resolution printed charts for two geometries, perpendicular rectangles with a common edge and aligned parallel rectangles are given as ".pdf" files on the CD.

References

1. Incropera. F.P. and DeWitt, D.P., *Fundamentals of Heat and Mass Transfer*, 4th Ed., Wiley, NY, 1996.

2. Çengel, Y. A., *Heat Transfer: A Practical Approach*, WCB McGraw-Hill, NY, 1998.

3. Holman, J.P., *Heat Transfer*, 8th Ed., McGraw-Hill, NY, 1997.

4. Ribando, R.J. and Weller, E.A., "The Verification of an Analytical Solution - An Important Engineering Lesson," *Journal of Engineering Education,* Vol. 88, No. 3, July 1999, pp. 281-283.

5. Modest, M.F., *Radiative Heat Transfer*, McGraw-Hill, NY, 1993.